THE PHILOSOPHY OF APPEARANCES

BOSTON STUDIES IN THE PHILOSOPHY OF SCIENCE

VOLUME 106

MIKLÓS ALMÁSI

THE PHILOSOPHY OF APPEARANCES

KLUWER ACADEMIC PUBLISHERS
DORDRECHT / BOSTON / LONDON

Library of Congress Cataloging-in-Publication Data
ISBN 90 277 2150 5

This is a translation of the original Hungarian, *A látszat valósága,* published by Magvető Könyvkiadó, Budapest

Translated by András Vitányi
Translation revised by N. Horton Smith, University of Nottingham

Distributors for the U.S.A. and Canada
Kluwer Academic Publishers,
101 Philip Drive, Norwell, MA 02061, U.S.A.

Distributors for Albania, Bulgaria, Chinese People's Republic, Cuba, Czechoslovakia, German Democratic Republic, Hungary, Korean People's Republic, Mongolia, Poland, Romania, the U.S.S.R., Vietnam, and Yugoslavia
Kultura Hungarian Foreign Trading Company,
P.O.B. 149, H-1389 Budapest, Hungary

Distributors for all remaining countries
Kluwer Academic Publishers Group,
P.O. Box 322, 3300 AH Dordrecht, Holland

Original title: *A látszat valósága*
First edition published in 1971 by Magvető Könyvkiadó, Budapest
First English edition published in 1989 by Akadémiai Kiadó, Budapest and Kluwer Academic Publishers

Kluwer Academic Publishers incorporates the publishing programmes of D. Reidel, Martinus Nijhoff, Dr W. Junk and MTP Press.

Printed in Hungary
by Szegedi Nyomda, Szeged

CONTENTS

Editorial note VII

Translator's note IX

Preface to the English edition XI

Chapter 1
ON THE ONTOLOGY OF 'PRESENTATION' 1

1.1. On the ontology of 'presentation' 1
1.2. The semantic range of the 'phenomenon'-category: the dialectic of 'being for us' 35
1.3. The semantic changes of a concept 64

Chapter 2
THE GENESIS OF VIRTUAL REALITY 83

2.1. The objectivity of illusions 83
2.2. Immediacy and the conditions allowing the occurrence of illusions 104
2.3. Object-making, everyday objectivity, averageness 124
2.4. The watershed of modern philosophy 150

Chapter 3
PRACTICAL CONSCIOUSNESS: SUBJECTIVE CONDITIONS FOR THE FORMATION OF ILLUSIONS 167

3.1. Practical consciousness: subjective conditions for the formation of illusions 167
3.2. Teleological thinking and the reflexivity of 'making' 168
3.3. The inverted world: final causality 175
3.4. The formalism of everyday consciousness 191
3.5. Ideology as quasi-reality 199
3.6. False consciousness and manipulation 227

Chapter 4
THE ILLUSORY FORMS OF EVERYDAY LIFE 241

4.1. The illusory forms of everyday life 241
4.2. Illusions of playfulness, irony and freedom 244
4.3. Ambiguity, misunderstanding and lying — the imperceptible 'quasi-reality' of everyday communication 261
4.4. Artistic illusions and the art of illusions 275

Index of names 287

Index of subjects 290

EDITORIAL NOTE

This ancient puzzle and common sense insight about appearance and reality reaches further with Miklós Almási's book than the reader may expect. He is, indeed, concerned about how misleading appearances are and continue to be, but the philosophical and common sense readers alike often misconceive the dynamics of untruths. Almási is blunt and lucid: what he will do is "to consider untruth changing into truth". And just as bluntly, can we bear the 'full truth'?

This essay in ontology and epistemology can teach us how to think about the believable and the unbelievable in these years of being trapped by appearances. Is it an essay in the philosophy of science? or in the philosophy of existence? But science too is part of the entrapment as well as partial source of knowledge of reality. We need not hesitate to welcome this book in its English version into the *Boston Studies in the Philosophy of Science.*

R. S. Cohen

TRANSLATOR'S NOTE

Just as translators of philosophies of reality must devise distinct equivalents for Hegel's *'Realität'* and *'Wirklichkeit'*, so the central concept of Miklós Almási's *Philosophy of Appearances* requires, as the author explains on page 37, the use of several terms roughly corresponding to the German *'Erscheinung'*. In this translation 'phenomenon' (in single quotation marks throughout the text) renders *'fenomén'* in the special sense which Miklós Almási defines, and the terms 'appearance' and 'phenomenon' render *'jelenség'*, the Hungarian vernacular word for phenomenon. 'Presentation' is borrowed from its maieutic context to render *'megjelenés'*, the act or event or manner of appearing. The Hungarian *'látszat'*, superficial or false appearance, is rendered by 'illusion'.

PREFACE TO THE ENGLISH EDITION

This book is about false appearances. About phenomena which may betray something about their origins and attempt to explain themselves, but can come up only with false explanations. (The sun 'rises' and 'goes down' — this is how we see it because this is how this phenomenon looks.) And our view of the phenomenon does not change much even though we know that all this is but false appearance. But this also applies to my money deposited in the bank which seemingly 'multiplies by itself', and this is the type of pretense the boss encounters who is 'adored' by his subordinates, who is praised apparently for his personal qualities, his cleverness and brilliant ideas but in fact really — at least largely — for his position and power.

The English noun 'appearance' may mean both 'looks' and 'false appearances', and so I am in trouble right from the beginning. The problem treated in my book is an experience we all share, and yet naming it is by no means easy. European philosophy (Bacon, Hegel) hit upon the concept which denotes this shared experience without ambivalence: whether we try to choose between the German *'Schein'* or *'Erscheinung'* or the French *'être'* and *'paraître'*, the latter also often used as a play on words. Appearance in the sense of *'Schein'* is not the individual error of the 'human factor', no optical illusion, but a pattern everybody sees in the same way and interprets in the same way — something that possesses interpersonal objectivity: it exists for us although it is not 'reality' in itself. Of course, it is not a matter of naming, of denotation, whether we can make a piece of our shared social experience subject to analysis or not. Still, let this only serve as a foretaste of the difficulties of translation.

Of course, the English edition was not only a linguistic and conceptual problem for me. Any writer considers a completed writing a lost child; as soon as he has finished his book the volume begins life on its own, becoming detached and alienated from the author. I am the same way with this book. I wrote it fifteen years ago and since then both the world and I myself have undergone substantial changes, and so it is no wonder that, although I have retained affection for it, I regard this writing with sharpened criticism.

The new edition presents an *a propos* for briefly describing this 'parent-child relationship'. To begin with, I still think that this book has a valid message for today just as I think it certainly had at the time of writing it. Very few writings have come out concerned with the modern processes of untruth changing into truth, although the resulting 'pseudo-reality' which is coming to engulf us must be regarded as a major and very dangerous latent disease of our times. Let us then pose the question, what has changed during the past fifteen years?

The first such comprehensive change conducive to appearances is the explosion of visual culture, the shift from written culture to pictorial culture, or, briefly — borrowing Adorno's term — the 'TV-effect'. This shift triggered significant changes in human consciousness. The movies, illustrated magazines and, chiefly, television and video formed in people the view and even ideology that things are in effect not very complicated, that what is visible, and even what is the secret motivating force behind happenings, is actually easy to investigate: appearances do not remain undetected, the inspector always wins. But parallel to this (to the situation of being viewers) there lives in us even a different consciousness, a healthily sceptical awareness which cannot be satisfied with the information, with conventional ready-made explanations for complex social, political or everyday situations. This 'other consciousness' looks everywhere for hidden reasons, secret motifs or more profound laws. All at once we have become more gullible and easier to influence as well as more suspicious and sceptical. We are living with a double consciousness.

For the time being these two kinds of consciousness coexist in us, but it is already plainly evident that under the impact of our coexistence with visual culture and chiefly the 'magic box' we are gradually coming to regard the happening on the screen as more reliable than our own first-hand experience, for that has not only the force of words but also the irrefutability of a document. And in the meantime we are gradually forgetting about our earlier practice of paying attention to secret motifs behind visible things and background information; we are no longer inclined to read between the lines and least of all are we interested in wanting to explore the secrets. It is possible that we have grown more accepting of false appearances and of coexistence with them. We take the pseudo-reality offered by the pictures more for granted, as more natural and more 'real'. Scepticism comes to have a much lower status for us: it is getting more difficult to distinguish between appearances and reality. The demand to know genuine reality usually wakes up in us only when we are faced with some life-and-death matter, but then we realize how poorly armed we have become for even making an attempt to explore the real features of truth.

Of course, the social sciences — including such new ones as the philosophy of language, communication-theory and philosophy of science — continue to investigate the differences between pseudo-reality and reality. However, the rapid spread of visual culture has permeated every aspect of our everyday thinking, our approach has changed to oversimplification and thinking in terms of clichés. In fact some researchers say that thinking in terms of pictures takes place according to a different model than our earlier cultural traditions have built. While the process of thinking that relied on reading was linear, 'visual thinking' is circular, which also means that our model of logic, and thinking based on thousands of years of reading, are also changing. The absolutism of logical and syntactic structures which has operated for thousands of years is breaking or at least fading. As a matter of fact, a circular or 'eidetic' approach may be more fallible in regard to false appearances; it does not require discursive analysis or the practice of falsification.

I doubt that this bitter hypothesis can be accepted but it is certainly not beyond my imagination to suppose that eyes used to visual culture require different logical structures. The question is whether this new attitude in seeing and thinking makes us more inclined to accept the everyday and social appearances of our times. I am afraid that the answer is yes.

Let me mention here also the commonplace occurrence that usually goes under the name of information dumping. This phenomenon also feeds our receptivity to appearances. It is certainly true enough that so great an amount of information is squeezed into us as to make us unable to digest the mass. The fact is, however, that the different information systems operate in such a way as to communicate largely mere facts, while the key information that would arrange the facts in a clear and comprehensive pattern no longer reaches us.

The first edition of this book treated the possibility of a sociological approach to these social pseudo-structures still with considerable scepticism, although it was one of its starting points, and in fact its methodological premiss, to translate this philosophical category into the language of sociology. Since then several important findings have been made in the field, especially in describing everyday appearances and prejudices, 'false consciousness' and the traps of communication which surround mankind. I am thinking of recent works of the trend which already was an important inspiration for my book at the time I was writing it. I have in mind the influence of Alfred Schütz's work on European and American sociology [A. Schütz: *Collected Papers,* Vols I—III, The Hague 1966; A. Schütz and T. Luckmann: *The Structure of the Life-World,* London 1973; A. Cicourel: *Cognitive Sociology,* New York 1973]. Also important was the publication of Jürgen Habermas's *The Theory of Communicative Action* [Vol. I, Boston 1983], which summarized the earlier findings of the Frankfurt School with respect to the theory of communicative action and produced new results particularly in the field of the actions and speech formations of everyday life. From the point of view of the theory of appearances, a new dimension was opened up with research on the subtle structure of prejudices. Earlier, both sociology and my own book considered prejudices a socially developed, though subjective, 'filter' for distorting reality. Habermas, relying on some important research findings, distinguishes between prejudices which assume negative and in fact aggressive forms in social and individual behaviour from a different type which bears one's commitment to earlier experiences and links to cultural traditions. In this distinction he goes on from Michael Polányi's *Tacit Knowledge* [London 1970] and Hans-Georg Gadamer's *Truth and Method* [London 1975]. He developed his view — controversial to this day — in a lengthy argument with Gadamer. Habermas says that without 'positive prejudices' it would be difficult to find our bearings in the world; in other words, there is no 'pure consciousness'. (This was the first lethal blow to German philosophy's *'Bewusstsein überhaupt'.*) Our views and approach to things are guided, without our guessing it, by this tacit understanding and the similarly tacit affinities to values which live in us. It is the first type of prejudice that makes one inclined to accept false appearances, but thinking in terms of traditions makes the critical judgment of some new interrelationship at least as difficult as negative prejudice does. It should be noted here that the views of Habermas and the current theory of false consciousness are still in the crossfire of continued discussions. [Cf. R. Geuss: *The Idea of a Critical Theory: Habermas and the Frankfurt School,* Cambridge etc. 1981; J. B. Thompson and D. Held (eds): *Habermas: Critical Debates,* Cambridge (Mass.) 1982.] These new lines of research, and chiefly the debates around them, have a bearing on the second part of my book, and I would certainly deal with their implications if I were writing this work today. To be sure, now I consider Habermas's reasoning the more convincing.

As I said, the basic thought of my book was to treat chiefly the existing 'ontological' pattern of appearances, in other words, to consider untruth changing into truth and the social derivations of pseudo-truths which everybody takes for granted; to work out how these pseudo-truths come into being in social and individual practice and experience. This is so although the second part of the book deals with false beliefs that have their roots in subjective and social consciousness which are regarded by given eras as actual laws on a virtually shared basis.

Still, now at some distance from the time when my book was written, I can already see that making a distinction between true values and pseudo-values, and altogether the inclusion of the theory of values, is regrettably missing from my work, though Talcott Parsons and the school he created turned this into a research area which received a great deal of attention in the 1970s and 1980s. Fads, the flourishing and fast fading of artistic trends, cannot be understood without an analysis of the domain of values. It is true enough that recent hermeneutics, which probes into this problematic, too, has been largely a European — or rather German — approach (Gadamer, Jauss, Bourdieu, etc.), but on the changing general approach to changing artistic and everyday values the reader will find a number of enduring questions in the works of Erich H. Gombrich and Nelson Goodman, and in Marx W. Wartofsky's 'Picture, Representation and the Understanding' [in: *Logic and Arts: Essays in Honor of Nelson Goodman,* Indianapolis—New York 1972]. What causes the problems in this area of research — and in any analysis in the field of the theory of artistic works or the history of taste — is that pseudo-values may often be preferred while genuine values may for long remain unappreciated. This is the *limen* where hermeneutical research stops.

The book tries to trace the analysis of ontological appearances — that is those which have come into existence — among other things from the 'reversing' structure of socio-economic processes. The fact is that these economic processes and their major phenomena assume a different countenance for lay and scientific public opinion in the 1980s. The great economic upsets of the 1970s — the oil crisis, the international monetary crisis and general indebtedness — did not make it easy to recognize the actual laws of these global processes. The Keynesian laws became just as questionable as the post-Keynesian or monetarist theories, not to speak of earlier theses of political economics. My book was still conceived in an optimistic period, and the screening out of the effective driving forces from the mesh of apparent interconnections is much more difficult in today's complicated situation, if possible at all. Who would have thought in the 1960s, for example, that the 'welfare society', which was then fast gaining ground, could suddenly lose ground and become obsolete as happened starting with the mid-seventies. The warnings from the Club of Rome seemed much too pessimistic — and still seem that way —

although today we sense that those global issues (diminishing raw material supplies, North-South tensions, shortages of water and food) will be the burning issues of the next few decades. And it still remains an open question what new appearances will be brought along with this new period which is now resting on a new and less firm theoretical ground, and what old, supposedly certain 'truth' it will change into appearances.

Of course, my book does not go into problems of economic policy. Its emphasis shifted with the change of the times only to suggest that, according to my present view, social appearances have by now become considerably less transparent than they were earlier, and the laws of national economics provide a less reliable aid than they did at the time. Nonetheless, I think that the idea in my book according to which the tendencies and laws which determine society, and the period generally, appear reversed and distorted on the surface of society, is still a tenable hypothesis. To this extent I still identify with the basic Marxist and Lukácsian position which determined the spirit and methodology of the book. As it is an open theory, I regard this tradition as capable of further development and of surviving even the difficult period of the present neo-conservative decline.

In conclusion let me say a few words about the general value assessment of everyday and social appearances. As I said, this book deals with 'false appearances', with misleading intellectual traps which deceive people. These traps constitute a danger both in thinking and in value orientation. I think I should supplement this one-sided looking for scapegoats: we might just as well speak about neutral or well-meaning appearances. In fact, findings in the sociology of knowledge have shown that appearances may often have a positive effect. I could say that it would be rather difficult to live without some life-sustaining appearances. It has been a long-standing problem for European cultural history whether we need the self-deceiving semi-truth of white lies or 'well-meaning' appearances. The answer of the Enlightenment had been a clear no. Nonetheless, already Ibsen had his doubts about the bearability of 'full truth'. His thoroughly decent, though naive, heroes, who wanted to enlighten friends and acquaintances about the deceptions inherent in their family and community life, usually met tragedy through the fact that those others, and then they themselves also, proved unable to bear the full weight of the truth thus exposed. And although in the philosophical sense I am completely on the side of a scientific endeavour to explore full reality, I am quite capable of understanding the purpose of the 'self-protective' appearances society has constructed.

The next question — and I suppose the reader will pose it himself — is whether there are any 'self-protective' and 'well-meaning' appearances of this kind, formations which actually express the yearning of this time and age for some kind of therapy. Personally I could only mention one of them, namely Ernst Bloch's by now famous *Prinzip Hoffnung*. Except that the principle of hope, although it can work as a balm, cannot be regarded as 'ontological appearance'. Hope has to be fought for by everyone on his own, just as belief in utopias. Without this, however, I can hardly imagine life.

What then, in my present judgment, is the attitude of man who has got used to appearances? Being entrapped without knowing it, a sense of living in pseudo-reality without awareness of our position. This is a strange state of affairs, though let me add that not even our life-world today consists entirely of appearances and less exclusively of manipulated conditions. There are still road signs in which we can believe. Nonetheless, this situation of being entrapped without knowing it makes us defenseless. We are surrounded by secrets, but we are not even aware of the fact that we need to count on coming up against them. In this situation we face an absurd choice: appearances offer themselves in such a palpable form that it is still more expedient to act on the basis of these appearances than to follow the laws of genuine truth, supposing that they are discoverable at all. This poses at the same time another question to both sociology and the philosophy of science.

Let us imagine that we are cast off on an island where the natives are playing the roles of a play that is unknown to us. All we, the new arrivals, notice is that no one behaves as would be natural, as would be demanded by their rational conditions. Here nobody gives himself away, everyone is 'playing' — and yet they understand each other perfectly. They perhaps know what this game is called. Now if we want to fit into this new island life, we have two choices open to us. If we began to behave in a natural way adjusting to our own standards and to the natural conditions, we would be immediately warned to conduct ourselves according to the roles we were given — notwithstanding the fact that these roles are foreign to us. Gradually and very slowly, after several failures, we will learn our roles. Sooner or later we will also act and behave according to the rules of the game and will experience success in the eyes of our companions. At the same time we will stumble against a number of conflicts presented by the 'real' conditions, by the natural environment which we have decided to ignore. On the other hand, if we take these natural stumbling blocks seriously, we will come into collision with the 'artificial' reality of the game played by the others. In this way, we are bound to develop a double consciousness in ourselves: in the 'play situation' we will accept our roles and the expectations of our companions as reality, whereas when confronted with the realities of nature and the realities of our own human standards we will automatically shift to a different, 'more realistic' pattern of behaviour. What is absurd in this double consciousness is that it offers several equally sad strategies. We can suppose that the 'island community' of the play is the 'true' reality and the natural world is sheer fantasy; we can regard the conflicts where the natural world intrudes on the reality of our game and upsets the shared illusion as exceptions; and finally we can play the game the others expect of us in a cynical fashion, but in that case we will also behave cynically when confronted by the real world because we will hardly know where one ends and the other begins. In other words, the internal boundary which sets off the two aspects of our double consciousness becomes less and less distinct and we will accept the 'reality of the game' as true, and we will force its laws into 'true reality'. We will keep shaping and reshaping the rules of the game until they will no longer be suitable for orientation in our interpretation of the real world.

Nor is this little travel tale in this introduction mere fiction. Innumerable historical examples could be cited to illustrate the bloody and blindfolded split consciousness of this 'play with reality' from the hypnotized masses of Hitlerite Germany to the ideologized cravings for pseudo-reality of the Stalinist period, and from the mass craze of religious or national hysteria to the make-believe world of 'subtle manipulation'.

One is sorry to say that awakening to realities seems to be possible only backwards in history, and even when our vision clears up in regard to history, some dimness is usually still left. Let us think only of all the 'secrets' of history which still have not found explanation. Our present consciousness is wobbling on the delicate balance between appearance and reality. What I would like is of course for this book to tip the balance toward a sense of reality. It is up to the reader to decide how successful it has been.

Finally, I would like to say thanks to the friends who were ready to give their advice and assistance while I was writing the book, to express my gratitude to Professor Emeritus Kurt H. Wolff, of Brandeis University, to Professor Robert S. Cohen, of Boston University and, of course, to the late György Lukács, who guided me in my work for several decades and to whose memory I dedicate this book.

CHAPTER 1

ON THE ONTOLOGY OF 'PRESENTATION'

1.1. ON THE ONTOLOGY OF 'PRESENTATION'

We might begin our discussion by analyzing the dialectic of the structure of appearance as something which has a significance beyond itself and which is itself a larger whole. Nevertheless I think that it is when this sphere of thought is set in its broader context that the real questions will arise. Does the structure of appearances exist at all, and if it does then on what level? The problem was stated by Goethe in an epigrammatic form and with resonant sceptical overtones: *'Natur hat weder Kern noch Schale'* — nature has neither kernel nor shell, that is to say, it does not exist in layers, it is not resolved of its own accord, as a consequence of its ontological potentialities, into appearance and essence, but rather it is a dynamic totality which is simultaneously appearance and essence, illusoriness and plausibility. Whatever divisions can be distinguished in nature must be taken merely as the work of man, the result either of individual or historical praxis or of changing degrees of apperception. As against this view of natural philosophy (and of epistemology) and chronologically almost at the same time there developed the Hegelian concept, the first far-reaching experiment with the objective ontological theory of presentation, breaking through the limitations of the Kantian initiatives. It stands to reason that it did not happen 'of its own accord' but rather by summing up, or to be more precise, by revealing the contradictions which were manifest in classical German philosophy, mainly in the intrinsic discrepancies of the Kantian system. In his *Critique of Pure Reason* [1781] he still conceived of phenomena only in subjectivistic terms, whereas in his *Critique of Practical Reason* [1788], by virtue of the analyses of ethical and moral factors, he had to treat the phenomenal world, too, as being real. The contradiction between the two works emerged and became the theoretical basis for the Hegelian concept. Namely, Kant only surmised that phenomenal existence had something to do with human practice, and that this medium surrounding us is related through the intermediation of human praxis to the deeper strata which exist as 'things-in-themselves'. This is the case even when these phenomena, in themselves, isolated, are of less than full value, are ambiguous or illusory. This surmise remained of course merely a germ of thought, or even more a contradiction of the system, since Kant could not have even

2*

the remotest idea of the real and virtual forms of presentation, even less of the more complex categories of real untruth. But Hegel's theory of appearance borrows precisely from these diffident attempts, from the contradictions of the Kantian system. The *Phenomenology of Spirit* [1807] created a revolution in philosophy by trying to derive the reality of phenomenal existence from the combined history of the development of human praxis and cognition. Its intellectual greatness is measured not solely by the fact that it raised the phenomenal sphere to the level of objective existence, though, if it had done only that, it would thereby have been initiating a revolution fraught with many problems. I see the importance of the *Phenomenology* in the fact that it attempted to develop objectivity in connection with man's self-creating process. The 'essence' — the human race — is 'in the making', in the historical process of self-creation. It is not a 'kernel' to which the individualities and the actions of each man cling as a 'shell', but it is rather a general process in which human praxis partly 'provokes' nature (so that it will show aspects of 'presentation' to satisfy its requirements and demands) and partly induces human society to actively reshape itself over and over again. Then in this dual process ever new strata of the human race may come to light. In this self-creating process the main attention is given to the constant change which this 'created essence' undergoes and to the historicity of its course of presentation. In the *Phenomenology* Hegel makes no mention of false presentation, running in the wrong direction or actually heading for an impasse, not even as a revolutionary germ. Notwithstanding this limitation, including also the systematizing auxiliary concept of *'anima mundi',* the *Phenomenology* conceals a speculative seed that is relevant even today, namely the idea of the phenomenal sphere 'being made', of its being a praxis category. But if we pursue this train of thought further, however cursorily, we may easily reach the conclusion that for the young Hegel the phenomenal sphere represented the socio-historical forms of presentation and that, as far as natural philosophy was concerned, he could imagine the existence of this relation as a category at best in an inchoate form. Because, if the phenomenal sphere is 'being made', then this applies primarily to society, and it can be related to nature only in the sense of a material exchange with nature, as well as in the sense of cognitive contact. The so-called natural phenomena in which we have no 'active' participation and which only 'surprise', hinder or help us 'of their own accord' do not belong to this active and humanly conceptualized concept of appearance.

All this is, however, merely a conjecture; the *Phenomenology* does not yet deny the natural existence of the category of appearance. The basic idea of the work is 'presentation', the historical 'making' of the 'objectification of the spirit', i.e. the peculiarity according to which it comes into being and is articulated in the self-development of the human race.

In Hegel's subsequent works, however, this brilliant idea sinks into insignificance, to become a trite and rigid thesis in the *Larger Logic* where he feels that he has to explain neither the fact of 'presentation' nor its genesis. Here the appearance-category is reduced to an apodictical proposition: 'the essence must appear'[1] *(Das Wesen muß erscheinen),* writes Hegel at the beginning of the chapter on appearance in the *Logic,* and thereby he reduces presentation to a clear axiom that needs no further explanation. In this manner, however, the development of the objectivity of phenomena is also impaired. The *Logic* is concerned solely with 'true' forms of presentation, i.e. with appearances (phenomena) that express the essence, and it disposes of illusions — the structure of 'false' or pseudo-reality — merely with a few, albeit brilliant thoughts by regarding them as a transitional stage between immediacy and the voidness of mere existence on the one hand, and the appearance that expresses the essence on the other. Even more savage in its effect is the intellectual self-mutilation which completely separates the appearance-category from the practice of the human race and changes it, on the whole, into a concept of natural philosophy. The peculiarity of social phenomena — i.e., the different degrees of objectivity of natural and social phenomena — which was still to be found in an indefinable, implicit form in the *Phenomenology,* has here quite disappeared. The dialectical articulation of reality and its unity is the play of totalities — which we will meet as the sphere of the contradiction between appearance and essence; this, however, stands apart from the range of human praxis, and is related to it merely as a result of the structure of the *Logic,* wherein the problems pertinent to ontology are raised from the point of view of epistemology, and the answers, too, are given within this framework. But the practice of the knowing subject has brought with it such a restriction of the concept of praxis that the relation between appearance and essence has also become narrower.

[1] *Hegel's Science of Logic,* Vols 1—2, transl. W. H. Johnston and L. G. Struthers, London 1929.

It may be presumed that Goethe's chagrined paradox (it is typical that he called it 'Ultimatum'), this protest against disrupting the totality of nature in this manner, was the result of his objection to this reduced formula. And as an exasperated protest — it is biassed itself. But the problem has been announced: does the category of appearance and essence, as a 'relational determinateness', apply to nature? Does only an inarticulated unity existing in a different kind of articulation exist in nature? Goethe's remonstrance can still be parried on the philosophical plane since the next lines of the 'Ultimatum' plainly show that, as against the disanthropomorphizing tendencies of the natural sciences, he is in fact defending the domain of art, which creates and cultivates nature as the extended totality of man which art has created; but he does so in opposition to the arguments of science and philosophy. The last lines of the poem, "Is not the kernel of nature in the hearts of man?", aptly indicate his mind which is fundamentally artistic, which anthropomorphizes, or if need be, re-anthropomorphizes nature. Goethe, the natural scientist, presumably did not consider the essence-constituting role of the 'human heart' as a scientific category. As a metaphor, however, it affords an amazingly accurate description of essence as a category of praxis. For something becomes essential or unessential (contingent, and so phenomenal, having the nature of a shell), only by way of man's given historical course of action and opportunities, i.e. through his praxis. It is essence which promotes or demotes particular relevancies, establishes orders of preference etc. This is the ontological hypothesis which the 'human heart' embodies as a poetic idea.

These two different views of appearance entered into competition with each other with unequal philosophical weapons. Behind one there were the results of the mature Hegel and of classical German philosophy, whereas the other remained at best a surmise or at any rate an artistic principle which had not been thoroughly thought out and had not ripened into a philosophical programme. It became a programme and a turning-point in philosophy with the writings of the young Marx. Here there was formulated with classical precision and pungency the principle of the relatedness of natural phenomena to man and to praxis as well as the basically — though not exclusively — socially oriented ontological structure of the appearance-category.

The relationship between appearance and essence is primarily a category of social life; it manifests itself in social praxis, i.e. in selection, in 'making', in the difference between the work of society as a whole and individual activity, and only the rudimentary forms of this relationship exist in nature. In the natural processes — particularly in the inorganic ones — the connection between appearance and essence is a relational category. In one direction it functions, it presents itself in some material connection, whereas in other directions it assumes a different aspect of appearance and essence and, in this respect, it is a category of rather relative validity. But in society it is characterized not only by permanence and relative variability but also by a high degree of man-created objectivity. In short: the totalities in nature can be disjoined through the category of appearance and essence only to the point of relativeness; on the other hand, this articulation is one of the foremost principles in the life of society.

Before we proceed, however, let me add a personal remark at this point. Our subject as well as the range of questions to be analyzed have to do with the origin of illusions and their functional structure. In this respect, the ontological structure of the relation between appearance and essence is in fact merely a problem of secondary importance and is introduced only as a measure of objectivity. But it is precisely in its capacity as an ontological yardstick that I have found the structure of this pair of concepts inconclusive, not only in Hegel but also in the later, modern conception of Marxism. With this end in view — though I felt I had to raise the problem — the forthcoming argumentations will not make claim to settling this contradiction, partly because I feel that my own familiarity with the exact sciences is less than sufficient, and partly because my knowledge of ontology and phenomenology is too superficial to take a strong stand on the pro and con arguments of this problem. My plan is that this study should be continued in the future with the exploration of the entirety of the Marxist theory of appearance. Under the circumstances, i.e. when we are dealing with such an overall theory of appearance, or, in other words, with the reconstruction of the Marxist phenomenology (if I am permitted to use this term), this range of problems, too, must be thoroughly thought out, and a decision must be made between the answers at present available. I therefore leave this partial problem with the reader in an unresolved form.

Let us now return to analyze the Marxist solution.

In the *Economic and Philosophic Manuscripts* Marx writes that natural phenomena have no immediate existence for man, that they are created only in the course of their social and practical transposal, through their 'being created', being made. "Neither objective nor subjective nature is immediately present in a form adequate to the human being. And as everything natural must come into being, so man also has his process of origin in history. But for him history is a conscious process, and hence one which consciously supersedes itself. History is the true natural history of man." [K. Marx: *Economic and Philosophic Manuscripts,* transl. R. Livingstone and G. Benton, London 1975, p. 391.] And elsewhere: "But nature too, taken abstractly, for itself, and fixed in its separation from man, is nothing for man." [*Ibid.*, p. 398.] These observations, as well as the categories 'form of presentation', 'tendency', 'inverted picturing' which have been developed in the course of analyzing the socio-economic structure indicate that not only in his juvenilia but also in his entire oeuvre Marx maintained the possibility of the conceptual solution according to which the relationship of appearance, essence and illusion in its most highly refined form is an ontological category of society, whilst as regards nature it is merely a category of praxis, and therefore an epistemological concept: the outgrowth of the expansion of practice, of the progress of cognition. From this point of view I consider highly important his above-quoted remark according to which history is the true natural history of man. He was to return to this idea also in his subsequent, more mature works (e.g. in *The German Ideology*) insofar as he regarded history as the only science, comprehending all other branches of science, even those disciplines which had earlier been or were even then regarded merely as of a fact-describing nature. This observation defines not only the fact of man's social development, that of the historicity of his 'species-being', but it also refers to the fact that nature, which reveals itself for man in increasingly deeper layers and hence shows manifoldly its phenomenal abundance, is a historical process itself, in a dual sense. On the one hand, nature also has a history of its own, and, on the other hand, the humanization of its system of phenomena and so its process of changing into a phenomenon 'for us' is also a function of this historical praxis, a consequence of this interaction.

Are we then justified in our allegation that according to this approach the dialectic of appearance and essence could be the category of social praxis, whilst its significance in terms of natural philosophy figures only in an epistemological and not in an ontological sense? Though the odds are in favour of giving an affirmative answer, I still will not do so, because — as I mentioned before — I think that the problem has not been resolved. Marx, however, in his subsequent course of development takes scrupulously good care to treat the conceptual variations of the appearance-category (such as 'form of presentation', 'tendency', 'mode of presentation' etc.) as concepts pertaining to socio-ontology, or when he does not do so, he at least distinguishes them as such. (Thus, for example, in both *Capital* and the *Theories of Surplus Value* he frequently underlines the ontological difference between, on the one hand, the forms of presentation and essential relations and, on the other hand, the distorted phenomena that come into being in people's minds in the course of reflection, i.e., in socio-ideological picturing, stressing that in reality the former appear factually indeed 'otherwise'.) These two different kinds of use of the concept can always be distinguished in Marx's writings since he uses very precise wordings to render perceptible the differences between the ontological and epistemological functions.[2] Insofar as he refers to examples of natural phenomena, they are either incidental or contingent on the historical movement of social praxis. In this respect, it is likely that Marx persisted all along in the socio-ontological interpretation of the appearance-category, whilst at the same time he treated its role in nature with scepticism, or at least regarded it predominantly as an ancillary concept assisting the advancement of cognition and so he presumed it to be of an epistemological character.

The likelihood of this assumption is confirmed by the continued progress of Marxism. No homogeneous ontological terminology developed either with Engels or with Lenin as regards this pair of categories. This 'reflectional determinateness' figures in at least two different conceptions. It is on the one hand applied mostly to nature as an epistemological category, insofar as it is supposed to indicate the changes in the progress of science, the changing of the hidden into the revealed. Though in Engels's investigations into the dialectic of nature there is constant evidence of an attempt to break through these limits (which have after all their origin in Hegel) and to apply the pair of categories as an ontological, historical category *vis-à-vis* the natural processes, this attempt, however, meets only with limited success. For example, in one place he explicitly sets forth the ontological absurdity of this attempt. In his *Dialectics of Nature* he quotes Hegel, writing as follows: Hegel himself states the true nature of essence-related definitions, in the *Enzyklopädie* [I, § 111., Supplement], namely that in essence everything is relative (e.g., 'positive' and 'negative', which have sense only in relation to one another and not as things-in-themselves). In other words, Engels, who discussed natural philosophy a great deal, surmises that the concept of essence can be only a relative category in the totality of nature, in the

[2] It is noteworthy that this differentiation is not so sharp yet in the *Rough Draft [Grundrisse]* where the forms of presentation are a great deal more strongly ontologized.

heterogeneous unity of relations; it can be valid only in one certain relation which is, however, overlapped and coloured, and hence made into something else by other relations. But it is consciousness which raises this single relation to make of it a subject of analysis, even if it has an ontological basis in its origin, and it is consciousness which recognizes the ontological relations prevailing here. Such relativity of the essential relations may account for the epistemological quality of the natural appearance-category (the appearance-category applied to nature).

Consequently, when trying to grasp the category of appearance and essence mainly as the objective configuration of motion of a higher form of motion, as a social concept connected with man's conscious practice, we are not after the conceptual discontinuation of the 'dialectic' of natural processes. We only want to refer to the fact that at a lower level of development some dialectical categories either do not occur at all (e.g., reproduction of the inorganic world), or occur only in a primitive or relative form of development. In our opinion, this applies also to the objectivity of the dialectic of appearance and essence. As an objective category, its real development occurs in society; its occurrence in nature is either relative (taking effect in single, individual relations), or is, for the most part, of an epistemological quality.

Lenin, too, in the main line of his Hegel studies, makes a point of focussing on the ontological core, but in some of his remarks (concerning particularly the description of the phenomenal sphere) he is inclined to treat this conceptual relation as the stages of cognition, insofar as it is applied to nature. So, e.g., commenting upon Hegel's formulation of natural laws as the 'untroubled contents of appearances', he writes that law and essence are homogeneous, or, more precisely, single-step concepts of the same rank which give expression to the absorption of human cognition in the phenomena, the world etc. The emphasis is thus put on the gradualness of cognition, and it remains open to doubt if indeed this articulation exists at all in an ontological sense, i.e., without the presence and contribution of the subject, and if so, then in what set of relations. Of course, these thoughts of Lenin's can be understood as arising from the philosophical debates of the period. Even after writing the *Empirio-Criticism* in 1908, Lenin still held it highly important to build up the epistemology of materialism, and so in the course of studying Hegel he steadily sought and elaborated arguments for use in this polemic. But dogmatism, instead of accepting the entirety of Lenin's system of ideas as well as his approach of polemicizing at times and proposing questions even to himself, tended to canonize individual quotations, a tendency which resulted in the fact that the epistemological view of the relationship between appearance and essence first became one-sidedly confirmed in the general consciousness, and then was directly identified with a primitive theory of ontology, namely that the same categories of appearance prevail in both nature and society.

But this epistemological description gives rise to an irresolvable contradiction. The essential and the phenomenal inevitably acquire a certain dualism of value: they become value categories. The essential will be the more important of the two, for after all it is that on which the phenomenal world depends, whilst the sphere of phenomena seems to blend with the unessential, with the less valuable. This is also due to the influence of the Hegelian tradition. In the *Logic* this dualism of value is the starting-point of the differences between illusion, appearance and essence. The second volume — with chapters introducing the theory of appearance — begins with an analysis of the difference between the essential and the unessential, developing from them first the category of illusion and then that of appearance. [Cf. *Science of Logic,* Vol. 2, Part I, Chap. 1.] But if we are now talking about a difference in value, this is something that can be established only by human praxis: it is only for man that something may be of more or less importance. For nature, as a totality, no such order of importance (hierarchy of values) exists, irrespective of the sort of articulation that we presume to be in it; with the supposition of such an order of values every materialistic ontology would be upset and a teleological or a Platonic false ontology would come into being.

Hence there is no doubt that in both Engels's and Lenin's views the problem of appearance figured mostly as a category of natural philosophy, but the consistent practice of this mode of application in philosophy would itself be confronted with a contradiction. It would become clear that we needed to return again to the concept of praxis — to assume the hierarchy of values as our premiss — and that we could not arrive at a 'noumenal' ontological structure.

Now let us see how this relation of presentation takes effect as the ontological category of social praxis. Not even here does the epistemological quality disappear completely, for indeed people consciously form the social processes, make and shape their lives, and in this respect they are more or less consciously subject to the world which they created but which is opposed to them. They are as liable to be influenced by action based on error, on false consciousness, on the illusory as by the gesture of an answer to real phenomena. Cognition figures as a factor in this conception, too, in evidence of the fact that the element of subjectivity cannot be completely eliminated either from the appearance-category or from the concept of essence, and, consequently, that this pair of concepts cannot be ontologized in an absolute sense. But here the chief constitutive factor of the relation of presentation is after all not the progress of the scope of mankind's knowledge but rather the reality created by social praxis itself, a reality in which the 'alien world' which presents itself for man and the objective fact of human action are, as phenomena, parts of one and the same structure, of one and the same social process.

The general movement in society is created by the 'team-play' of individuals who have specific aims and interests and who are furnished with exact 'initial conditions' determined by the position they occupy in the division of labour, and by their social stratum, but the result of this 'team-play' is the already 'complete' although dynamic totality, i.e., the ontological basis. At the same time, as an active subject the individual person always faces only a partial phenomenon, one which directly presents itself for him, a shred of the reality which he himself 'has made': the distorted projection of the Whole. (This projected image is distorted still further by a countless number of factors, by the various configurations of social consciousness, from ideology to the common form of consciousness.) The 'presentation' is nothing else but the transformation of the general movement into the world of individuals: it is the peculiar, distorted and homogenized configuration of presentation as the various effects of the general movement penetrate into some sphere of life, that which individuals grasp as an objective picture. The fact of the transformation, the 'picturing' ('mapping') of the general movement onto the socially delimited individual 'situations' is not a function of the subjective activity of cognition but is an ontological process. Consciousness has nothing to do with the sort of structure that the individual person faces in the course of his daily work, the events surrounding him, his family life or his tribulations at the office. Consciousness is a subjective component which may at best have a bearing on man's life only when he can understand and utilize in his own practice some of the mass of phenomena which offer themselves 'for him' and which are transposed onto his own sphere of life from the general movement.

What we have here is even the dialectic of the circumstance that the same person who as recipient, as 'knowing subject', faces merely a segment of the general movement is naturally the creator not only of this segment but also shapes this general movement through various structures. (We shall see in the next chapters how this dual role manifests itself on different levels in, for example, the process of competition, where the capitalist depends on price fluctuations — a shred of phenomenon — and at the same time he is unconsciously one of the constituents of the general movement in his active life.)

Let us examine an elementary example of the way in which this ontological relationship develops in social praxis, how the respective relational categories of appearance and essence differ from and are intertwined with one another. Social reality consists of the existence within one another of various totalities, classes, strata and groups in the division of labour, interest alliances, social cells etc., and it feeds on their interaction. But these totalities also include new elements themselves, so the connection between the larger units develops simultaneously through the relation between the individual constitutive factors, i.e., through the individuals' immediate and active connections with one another, and the transposal of the general movement that they have created themselves. The private connections and the deeds of the individual acting persons are phenomena, but the relations of the larger units that are based on these phenomena and which at the same time determine the latter's individual movements constitute the sphere of essence. Social reality is thus both the difference and the unity, or more exactly, the simultaneous existence, of these two planes. Let us illustrate this relationship with the example of the road traffic in a small town. The set of a specific number of cars must 'live within' the road-network of this small town in which they have to find those lawlike possibilities by means of which they can mutually affect, expediate or hinder their respective movements. For the sake of simplicity we have chosen an example in which we have to consider only one of the two totalities existing within one another, namely the set of cars. In what way and for what reason are the movements of the individual cars (as phenomena) separated from the general movement, and how do the two movements — the phenomenal and the essential — still exist simultaneously? If we start out from the totality of the vehicles it is easy to realize that the pace of the traffic, its possible speed, the standard of habits developed and the average number of accidents depend on the number of cars. The larger the number of cars is, the slower the pace of the traffic will be, up to the point when, reaching saturation point, a traffic jam sets in: the individual cars will prevent one another from moving along, and collisions will become more frequent. At this precise instant there becomes manifest the difference and the unity between the individual movement and the general movement; each driver has his own particular way of driving (cars of different capacity), with a different purpose and proficiency, so each case of congestion will be brought on by different (accidental and individual) causes. Yet, in these 'jams' the individual vehicles come up against all the others: it is the total traffic which blocks the road for the individual vehicles. Though their individual character differs from the behaviour of the totality, here they still encounter one another. By virtue of their ability and their skill in recognizing the prevailing conditions and also depending on the accelerating power of their respective cars, the individual drivers could perhaps evade these 'jams', but there is no way of doing so because the scope of their possible movement is determined by a larger unit ('all the other vehicles') which is seemingly unrelated to them. (It forces them to slow down, accelerate or cause an accident.) The congestion is caused by the accidental — phenomenal (apparent) — clumsiness of a single car or two, but the sum total of vehicles is also present in this incident since the cars create for one another that larger unit which now imposes its power on them as individuals. The reason why two or three vehicles 'by chance' block the traffic — thus getting into a situation in which they are unable to proceed — is due to the intermediacy of all the other cars. Their individual movement constitutes the totality of vehicular traffic, but this totality will also react on them, affecting their ability to move, irrespective of their intentions and individual capabilities. Only the individual cars meet, evade or collide with one another in the street, but this individual and adventitious movement is determined by the quantity of cars, by the sum total of vehicles. Just as an increase of pressure affects the kinetic motion of gas molecules so the increase in the number of cars affects the entire totality, and this extrinsic effect is 'distributed' by the random individual movements through 'chance events': new traffic rules, accidents, jams and collisions. Hence the relational order between the movements of the individual cars and the entirety of the traffic is established reciprocally: one is determined by the other, and the phenomena of traffic are created by their existence in one another.

Still, the two planes are separated. The interrelationship of the individual constituents and the connection of the totalities are at variance with one another and through this difference they constitute a unity. The phenomenal sphere is the medium of substantive relationships (represented in the example by the statistical laws of traffic movement), and in this respect both the difference and the unity exist 'on the same level' while constituting two different determinatenesses. Hegel formulated the difference and the unity, the separation and identity of appearance and essence as follows: "For (considered more precisely) determinate essential content is not only determinate in general, but as the essential part of

Appearance, it is complete determinateness: One and the Other. In Appearance each of these has its persistence in the other in such a manner that it is also only in the other's non-persistence. This contradiction cancels itself, and its intro-Reflection is the identity of the persistence of each: the positedness of the one is the positedness of the other also. They constitute one persistence, and at the same time are a various and mutually indifferent content." [Hegel: *Science of Logic, Op. cit.,* p. 134.] So the world is not divided into two distinct spheres, 'they constitute one persistence' in the sense that appearance exists just as the essential relation does. Yet, it remains set apart, for the individual phenomena carry only a fraction of the mode of existence of the general movement, of totality. As phenomena, the individual jams, collisions and accidents follow from the 'distribution' of the increase in the number of vehicles, of the motion of totality, but at the same time they have some other — individual — particularities as well (e.g., the driver's individual reflex and his experiences also have a bearing on which particular person will be involved in the statistically 'prescribed' accident).

The world of society has indeed neither 'surface' nor 'nucleus'; existence simultaneously comprises both. The substantive relationship is nevertheless separated within it, the general movement from the contingency of the individual movement, but in such a way that it is through this difference that the above-mentioned relationship realizes the totality of its existence. This is the law of transcendency: essence is 'beyond' the individual movements since the individuals while performing their self-movement create in combination with all the others the quality of movement of the larger unit which rules over them. 'Everything else', that is, totality, therefore exists in a relation other than that in which the individual quality, which is its constituent, does. By analyzing the movement of capital, Marx demonstrates that the general movement may even acquire some relative independence. He says in the *Grundrisse* that capital in its generality, in its dissimilarity from the particular, real capital, is a configuration which has true existence itself. In the course of competition individual capital comes into collision precisely with the configuration which exists in this general form. But at the same time Marx also points out that this process of becoming independent takes effect only as a relational category, namely the separation of the general movement becomes an independent configuration with true existence only in relation to its own constituents. The general — essential — mode of existence cannot be detached from the individuality which constitutes it, but in relation to that individuality it still has a certain form of independence.

What does all this mean from the point of view of social phenomena, from that of socio-ontological 'presentation'?

What it means first of all is that phenomena (in this case the individual actions, the potentialities manifesting themselves in a consciousness which intends to understand the world) convey the interactions of totalities only in fractions. By the time this process enters into the individual's conscious orientation it assumes in its other ontological structure, too, merely a relational configuration which, however, embraces the totality of the general movement, only in a different form, in a configuration pictured onto this single relationship. We might call this relationship the law of the transformation of appearance. It is not consciousness, individual cognition, that moulds and distorts reality into partial phenomena or indeed into illusory forms of presentation; rather, it is the general movement itself which, by means of the contact between the many different kinds of totalities, through their immanence in and connection with one another, is changed into a relational form, the compass of which is determined by the given individual's conditions of existence (circumstances of life, work, or scope of existence). The general movement is pictured on this given compass, on the 'local form of motion' which functions here, and which is also made by the individual — in accordance with its requirements. Hence the point under discussion is not that cognition dwells on single components only out of the social totality or that the progress of our scientific analyses can approach the great whole only step by step. The basis of the ontological structure is always the totality, but its concrete existence, developing from and reacting upon the individuals' active existence, always blossoms only out of transposals.

The fractional parts of the released phenomena and the relationships receiving them are in contact with one another merely on a minute surface in any interaction, for the time being independently of the individual reflections. It is through the 'transposal' of the money-market that credit and currency are in contact with the general movement of capital, and the connection between these two totalities means at the same time that the general movement of capital will affect the laws of finance only in a form of transformation. The given motion of the aggregate capital presents itself for the credit system *ab ovo* in a metamorphosed form. Here presentation still has an objective structure, and it is only after this that consciousness of the individuals comes into play, either discerning this minute mode of contact or remaining within the compass of the given, transformed picture.

At the same time, the transformation of appearances (phenomena) does not preclude the success of totality in the given, 'limited' setting of the presentation, but it only allows totality the opportunity to assert itself in an altered form. That is, the general movement participates in the individuals' or groups' segmentary life, and it does so not only as a relational, fractional existence. The dynamic connection between the great totality and the partial fields can take effect only when under the impact of the relationship the entirety of the original totality is transposed on this single relation and exerts its influence, transformed, but in its transformed state condensing its entire self in this single — or few — relation(s). (In the fluctuation of the rate of interest the dynamics of the different components of capital as a whole exert their influence in a 'reflected form', but the trend of commercial or industrial capital and of the average rate of profit cannot be eliminated from this effect, nor can the 'determining agent' — which after all has an influence over the size of the rate of interest — be reduced to a single constituent. It is the general movement itself which participates with all of its elements in this creative role, albeit in an altered form, projected on a single plane.) The concept of transformation embodies the following relationship: when in three-dimensional geometry a solid body is transposed from one plane to another and its entire image manifests itself in a new projection but, in accordance with the new plane, in a 'deformed' visual angle giving prominence to one particular aspect, so here, too the same 'picturing' occurs, only in a simpler form, two-dimensional and without quality.

Everyday social relations also operate with this transformation and thus create phenomena. When we have to act in our individual social functions (roles), e.g. 'as' a teacher or 'as' a father, these functions will have a true effect only when our entire personality is focussed on this single plane. The pupils for example will immediately notice the difference between the teacher who 'only delivers the lecture' and the one who is passionately fond of teaching, though their contact with their instructor is only through the 'filter' of the teacher-pupil relationship and they have little or no knowledge of their teacher's other human qualities which manifest themselves in other fields. Still, this 'single relationship' is sufficient for them to have a feeling as to what sort of a person this teacher is, because the completeness of the individuality is also 'pictured' on this relationship. The teacher's activity becomes an objective phenomenon 'as such'; partly it will conjure up the totality behind it, but as a transformed configuration it will show it 'in a different way', concealing the profusion of his totality.

The 'picturing' of totality is not an unfamiliar notion, it is one of the basic phenomena of artistic reflection, too. In his *Aesthetics* Lukács searches into the artistic consequences of this transposition while analyzing the 'homogeneous medium': for instance, one makes contact with the medium of the artistic creation with one's eyes when approaching a painting and with one's ears when enjoying music, yet through this single relationship one can build up for oneself the totality implied in the work and at the same time enters into it as a whole person through this single surface of contact. And the medium of the work itself — e.g. acoustic material — can thus convey the fullness of reality. Not only does the dimension of sound have an effect in a piece of music but also the multi-directional relation of the world it evokes. Here, too, the immediacy of the one-sidedness of the relation mediates the dialectic of the totality: in the contact of the parts the totalities are condensed and reproduced. Accordingly, in the social presentation the phenomenon is 'preformed' in its objective quality. It develops in advance (prior to the subjective reflection) its particular, transformed configuration in which the components of the general movement and the conditions of existence of the individual who with his own immediate activity helps to compose that general movement jointly shape the form of presentation which the individual must understand, in which he must orientate himself, and which will become the starting-point for his action.

The second consequence is the contradiction of the objectivity of the social phenomenal sphere, meaning the entwining of objectivity with the subjective factors produced by individuals. In other words, it means the solution of the long-standing problem in the history of philosophy known as the unity of subject and object, a unity which takes effect through various contradictions. This socio-ontological discovery of Marx's resolves the strict dichotomy which dominates the history of philosophy (apart from Spinoza and Leibniz, the great experimentalists of the 17th century), a dichotomy which is the fetishized separation of the subject — who has been reduced to a cognitive conscience — from a rigidly contrasted natural and social objectification. According to this view, whose best known and most satisfactory summary was given by Kant, consciousness merely contemplates reality, and praxis has no place in this contemplative attitude, so that practical considerations, whether moral or social, can be treated only in a separate philosophical system which conflicts in numerous respects with the basic system founded on this intuition. (This is one reason, among others, for the antinomy between the *Critique of Pure Reason* and the *Critique of Practical Reason*.)

And indeed, the concept of unknowableness — of noumenon — also results largely from this purely contemplative relationship, from the absolute separation of subject from object. Man-created social reality is cognizable, or at least in principle there is nothing to prevent it, whereas it is well-nigh impossible to find contact with the thing which presents itself for man as an alien world, or indeed as a wall towering above him, in the form of objectivity. At the same time, it is also obvious that the Kantian concept is only a synthesis of the scientific trend originated by Descartes, a trend which took nature to be the model of the world of science and of ontology, and which either regarded social existence as a natural basis or ranked it within the domain of natural phenomena. The great discoveries of this view were the outgrowths of the scientific revolution, but at the same time they also facilitated the praxis-centred understanding of man's social existence or the social totality. This turn towards practice begins with Kant's critique. As Lukács makes clear already in his *History and Class Consciousness,* it starts with Fichte's system and achieves relative completion with Hegel. The term 'relative' is to be understood literally, since Hegel, too, is unable to grasp the dialectical unity of subject and object. He takes only the first step towards eliminating this strict dichotomy. It is a matter of common knowledge that the basis of his system is the unity of subject and object: the *anima mundi* sets out from this unity, and after making its rounds through the various configurations of nature and history, returns there. This mystified and dogmatic framework, however, conceals a brilliant root idea, namely that the true unity of subject and object may be realized in the historical course of events, through a process in which individual people make their own history and they grasp and reflect this 'making' — taken in a historical sense — on various levels of the forms of consciousness. Here subject and object do not face one another in a solely epistemological abstraction, but the individual, through his consciousness and his active existence, is simultaneously the maker and the object of the process that he will later grasp; he relates to it simultaneously as a creator (as part of the general process) and as a knowing subject (as one who understands the process, being in possession of particular forms of consciousness).

With Hegel this concept cannot develop with all that follows from it because the system's requirement for a conclusion (the 'end of history') as well as the abstract praxis category (taken only in an intellectual sense) makes it impossible for this unity as an infinite process to be conceptualized as an interaction between the general movement of society and the individual shaping of history. This is possible only on the basis of a materialistic ontology of society. This interpretation is achieved by Marx, first as the correction of the Hegelian dialectic in the *Economic and Philosophic Manuscripts* and *The Holy Family,* and then in the socio-ontological system of the *Grundrisse* and *Capital.* To begin with, Marx reverses the preconception of 17th-century origin which regarded the exact sciences as the methodological model for philosophical investigation, and history as the only basic science.[3]

Hence the objectivity of social movement, the dialectical structure of the great totality developing from individual praxis, becomes the new ontological model. This is where praxis becomes a category of existence, resulting in the resolution of the dilemma arising from the extremes of absolute identity or absolute opposition. Marx's entire fetish-theory is based on the contradiction between false objectification and false subjectification, a theory which of course postulates that the true dialectic of these two constituents is the source of his chain of ideas.[4]

We shall often return to this dialectical relationship, the connection whereby the general movement is compounded of individual (subjective) actions and yet achieves objectivity. At present, however, we can scrutinize only the significance that this root idea has in philosophy: the elimination of the merely epistemological attitude, as well as the weakening of the one-sided method of natural philosophy, and thereby the normal arrangement of the natural forms of motion.

At first glance, it can only be an idealistic system in which the identity of subject and object can be taken as the starting-point for describing an ontological structure. But only at first glance; in fact Lenin exposed this

[3] Marx recognizes nature only as a frame of existence of socio-historical man, as the 'background' and the scope of action of this existence: "The human essence of nature exists only for social man; for only here does nature exist for him as a bond with other men, as his existence for others and their existence for him, as the vital element of human reality; only here does it exist as the basis of his human existence." [*Economic and Philosophic Manuscripts, Op. cit.,* p. 349.] Then he further concretizes this idea for capitalist development: "Thus capital creates the bourgeois society, and the universal appropriation of nature as well as of the social bond itself by the members of society. Hence the great civilizing influence of capital; its production of a stage of society in comparison to which all earlier ones appear as mere local developments of humanity and as nature-idolatry. For the first time, nature becomes purely an object for humankind, purely a matter of utility: ceases to be recognized as a power for itself." [K. Marx: *Grundrisse,* transl. M. Nicolaus, New York 1973, pp. 409—410.]

[4] For example: "The communal or combined labour posited in this way — as activity and in the passive, objective form — is however at the same time posited as an other towards the really existing individual labour: as an alien objectivity (alien property) as well as an alien subjectivity (of capital)." [K. Marx: *Grundrisse, Op. cit.,* p. 470.]

apparency when — following Marx's lead — in his *Materialism and Empirio-Criticism* he pointed out that the contradiction between matter and consciousness is absolute only from an epistemological point of view and that in all other respects it is relative. And though *History and Class Consciousness,* Lukács's first large-scale essay in ontology, is only a germ of his *Ontology,* his purpose was to concretize and extend this Marxist-Leninist idea in philosophical terms. His objective was to grasp the category of existence as social praxis and thereby to break away from the mechanical, materialistic contemplation of existence regarded solely as 'real'. [Cf., e.g., his *History and Class Consciousness* and *Reification and the Consciousness of the Proletariat.*] He endeavoured to develop the Marxist meaning of the unity of subject and object and its consequences for philosophy, i.e., to develop a social and historical ontology. It is evident that this approach was a brilliant, but in many respects abstract, attempt to delineate an overall structure of ontology, the positive solutions of which are perfected in his subsequent explications, partially through the criticism and development of abstract and half-solved problems. Here there emerges for the first time the definite formulation of the philosophical programme of Marxism, by way of reconstructing Marx's theory. What we have here is the philosophical requirement that we understand social movement, the dialectic of individual existence, not through the natural forms of motion but rather the other way around. What we must see is that individual activity simultaneously creates and experiences the objectivity of socio-historical processes. "Only if the subject (consciousness, thought) were both producer and product of the dialectical process, only if, as a result, the subject moved in a self-created world of which it is the conscious form and only if the world imposed itself upon it in full objectivity, only then can the problem of dialectics, and with it the abolition of the antitheses of subject and object, thought and existence, freedom and necessity, be held to be solved." [Lukács: *History and Class Consciousness,* transl. R. Livingstone, London 1971, p. 142.] Therefore, even though Lukács subjected his earlier study to a sober and theoretically correct criticism, it has become a source of the ramifying phylogenesis of the modern development of philosophy. Not only Marxist but also Marxisant thinkers have drawn upon it a great deal. The additional concretization of the root idea was imperative for the philosophical renaissance of Marxism to be founded on Lukács's newly initiated line of thought, but with an ontological theory more nearly approaching the Marxist tradition in intellectual profundity. Both the Frankfurt School (Th. W. Adorno, Horkheimer, Marcuse) as well as

Sartre and Lucien Goldmann borrowed from this work, although this development took a rigid course, and deprived the idea of Marx and Lukács of its positive dynamics, and at times of its socially active revolutionary content.

Regardless of what sort of an abstract form the idea of the unity of subject and object assumed in this first attempt at ontology, it brought two significant discoveries which are still potentially influential even today. One of them improved on Marx's idea that social phenomena are formed in the praxis of individual persons through the transposal of this praxis. The numerous individual actions, blunders, interests etc. develop a certain tendency which is perhaps contrary to every agent's will. This general movement and its conformity to laws will necessarily be at variance with the individual acts and mainly with the way in which the actors understand the motion of totality. The general movement manifests itself for the individual in different transposals because the social conditions of their individual praxis form a screening device through which the general movement is transformed, and which also changes its realizable structure. Lukács regarded this dialectic as smooth and unhampered in the case of the proletarian consciousness. According to him, going beyond the bounds of immediacy is not only a necessary step, but also one which always ensues; it is a stepping-stone which virtually follows from a conscious combination with totality. This premiss did not prove to be true, and he found the correction through his additional explorations in ontology. The other feature of Lukács's thought is that the unity of subject and object (the attitude which creates phenomena and consciously understands them) is realized on various levels. As against Hegel's dogmatism, Lukács perceived that not even as a final product can absolute unity be assumed in this relation. What unity there is is realized at the various levels of existence and consciousness, fragmentarily, and in the abstract, on the scale of immediacy, and producing illusions. This unity becomes more complete on the level of totality grasped with the aid of successful conceptual machinery, while on the level of a conscious and efficient shaping of society, i.e. of revolutionary praxis, it becomes a true kinetic identity. This idea was for Lukács at that time merely an unanalyzed inference. In his later ontological research he focusses on the fact that the unity of subject and object has various levels. We, too, shall stress the importance of this notion in connection with the analysis of the objectivity of the categories of appearance and illusion. We shall see that the concept of 'phenomenon', the abstract and immediate appearance-category, which in the course of the history of philosophy has been interpreted in so many different ways, is in fact nothing else than the direct unity of subject and object, insofar as consciousness does not make allowance for the irradiation of subject-being (the effect of action and means, their distorting and formative influence on the object) on the object which consciousness has 'pictured'. Consciousness regards as objective that which includes a number of subjective elements, without being aware of these subjective distortions. This structure, the 'phenomenon', is thus only an elementary level of the unity of subject and object, and since it does not have its origin in totality, it is saturated with numerous illusory elements. (To revert to an earlier example, at the early stages of astronomical observations there were rings noticed round each star, due to the imperfect quality of the lenses. However, what could not be known was that this was not the objective property of the stars but that it was a distortion caused by the instruments, hence that it was the effect produced by the subject upon the object and interiorized by cognition. The subject knows an object that he also forms himself. In the 'phenomenon' which manifested itself in this manner the subject should also have observed himself, his own effect. Here immediate unity on an elementary level resulted in illusions.)

In our opinion, this relation, the unity of subject and object asserting itself through contradictions and embedded in praxis, is a theoretical basis for the appearance-theory. Even 'presentation', the 'phenomenon' which manifests itself for man, comprises the 'question' put to their environment by the acting individuals, and the 'answer' given by reality, that is to say, a unity of subject and object on an elementary level (an attempt by reality to form something) and the difference (the answer of reality which is 'other' than expected). Here already the process of becoming a phenomenon emerges from the sphere of existence created by practice. Were the subject facing the object merely as a passive recipient (as a knowing subject) as an extraneous observer rendered independent of his object, he would have recourse only to 'appearances' which offer themselves 'of their own accord'. But even in our connection with nature it is rather rare to find a datum of this kind which confronts us 'automatically' as a phenomenon. As a result of our activities, here, too, it is the oddity of 'reaction' and of 'transformation' or its cognizable, pliable or rejectable structure that we regard as a phenomenon. In society, on the other hand, by definition, the only kinds of phenomena that exist are those made by people classed within the various historical categories of social existence. Here there exists no phenomenon of the type which is at variance with this mutual relationship. Man is simultaneously the subject (discoverer and creator) of what is assuming the form of a phenomenon or process in the transposal of the general movement, and the object of other people's active effects changed into individual and social objectifications. *A priori* no phenomenon can exist other than the identity inherent in the unity and the difference of subject and object, and without this relation we cannot understand the objectivity of phenomenal existence. A society in which all members are merely observers of the phenomena of their collective life is a *contradictio in adiecto:* in this case there would be no phenomena, because these are nothing other than the interactional configurations (assuming independent forms in the course of the interactions) of their activities, and the 'picturings' of these configurations.

It is evident that even in this early ontological essay Lukács intended to rehabilitate a fundamental relation of philosophy, and even though at that time he achieved only an abstract solution, his endeavour was to bear fruit in the underlying idea of his *Ontology,* his clarification of the concepts of labour, praxis and teleology and his advanced elaboration of the fundamental ontological relations of society. Though from a sociological point of view *History and Class Consciousness* represents an abstract solution, we cannot say that this construction is no more than the artificial extension of classical German philosophy, a claim made by both his critics and the bourgeois continuers of the objectification-theory (e.g. the Frankfurt School — Adorno, Marcuse). Revolutionary praxis is as inconceivable without this dialectical relationship as the philosophical extension of Marx's theory of fetishization and objectification. That is why we shall take *History and Class Consciousness* as our starting-point as regards not only the questions of the appearance-theory (which are merely touched upon in the present study) but also the objectification of illusions.

The separation of the level of the general movement from that of individual praxis and the unity of these two levels are conducive to raising a third problem closely related to the former two, namely, that the objectivity of phenomena cannot exist without the identifying, employing and transforming role of consciousness. This subjectivity is not its creator, for we have seen that, according to the law of the transformation of appearance, the configurations of presentation reach the level of individual activity and the mode of existence in a preformed state. It is not consciousness which gives them their mode of existence, their special — transformed — quality, or the totality content inherent in them. However, it is only through consciousness that what presents itself in this way becomes an action, that is, a new phenomenon, one which continues to exert its irradiation and its influence. Only through this subjective circuitous route will the individual perception (or mistake) result in an individual action which is — in the entirety of the general movement of society — a phenomenon. And indeed, at this point the previously mentioned value aspect becomes relevant. Man's conscious action — the 'responsive man' (Lukács) — makes a choice between alternatives that are essential for man and ones that are less so, but this substantive selection is now only the result of a basic relationship (of the characteristic of phenomena that they 'demand' response) and the form of presentation of a course of action which builds and is steered by a greater totality. (The individual customer makes a careful choice as to which pair of shoes he ought to buy, and he finds 'substantial' differences as regards the exterior of the shoes, but the substantive laws of a greater totality will take effect through the statistical quota of the customers, the rising marketability of some type or the slackening demand for others.) The active influence manifests itself here simultaneously as perception, choice, action and 'motivatedness' controlled by society. It is a phenomenon — not for the acting individual but rather for the structural teamwork of society. It is both a passive and a constitutive element, but the perception of phenomena which manifest themselves immediately and demand, direct or restrain a response, and the active response to this perception — i.e. the origination of new groups of phenomena — can materialize only through this subjectivity, through its conscious reflexiveness. The conscious element cannot be eliminated from the sphere of social phenomena. But then how can we possibly treat the phenomenal sphere as an objective configuration, as an ontological category?

The answer is partly included in the question itself, and all that remains to be done is to make it explicit. Individual fate, choice, active response, in short, man as a 'choosing being', is not only an imaginary formula reflected by thought but, through his active influence, is also a part of objectivity. Man as chooser represents a factor of appearance which — besides becoming a phenomenon in reflection, in the power of receptive consciousness to 'observe' and to realize — also produces an objective influence upon the general movement of society. The collective action or controversy of a number of people and the activity of social groups form a set of phenomena that can significantly shape society. What we have here is no longer the subjective perception of phenomena, but the ontological configuration of presentation. We might call this configuration 'being for others'. Although this category comes from Hegel *(Sein für Andere),* in our view it figures in a wider sense, with an ontological structure and — as we shall see later — it helps us delineate the ontological differences between natural and social phenomena, and also helps us keep the appearance-category's natural philosophical content, while letting it speak for itself in the form of development adequate to the specific character of the various forms of motion.

What does the concept of 'being for others' mean? Its essential character is this: that which 'presents itself' rises to view not only and not mainly for consciousness, still less for the subject. For it to become a category of consciousness it must become first an efficient (active and action-provoking) phenomenon for the rest of the active individuals, institutions and elements of reality that constitute society. The act of consciousness which fixes its reflected, understood and consciously perceived object as a phenomenon is only a subcase of this 'being for others'. The hitherto established theories of appearance, in particular phenomenology, proceed merely from 'being for us', from its subjectivistically isolated form, and thereby they have remained set in an epistemological relation, in the 'short-circuited' connection between consciousness and object. This 'short-circuit' means that abstract consciousness, which is independent of practice, faces an 'object' that is likewise exposed to no formative or other influence. Consciousness can therefore observe only that what the 'object' allows it to observe, only the side that the object turns toward the knowing subject or the side that it turns towards social praxis unknowingly. But this short-circuit also means that the individual is unable to decide for himself whether that which presents itself for him is only appearance or a true relationship. He has at his disposal merely a subjectivistic impression. He cannot measure the totality of the object and its relationships against the universality of experience and praxis. Hence 'being for us' displays an abstract and subjectivistic configuration. In this respect, it is the prime source of every illusion. However, with 'being for others' this 'short-circuit' disappears, since perception is assigned only a secondary role, the primary one being the effect for the other, the influencing of the active existence of the individual or of the 'others', to which one's answer most often is that 'he does it but does not know it'. Hence 'being for others' premises a manifold objective system of interactions of which the consciousness of the subject is but one — though no doubt essential — constituent, but the frame of which is the 'effect irradiated' and the 'receptive' answer, that is, praxis. This is the manner in which the criterion for the objectivity of the phenomenal sphere is created. It exists not only in its being for me (on this account it may still be a false appearance, a fancy of the mind or an optical illusion), but also for everybody else. It exists objectively in the various configurations of the different forms of activities in comparison with which the understanding of these forms, the conscious evaluation of the effects, the explanation of their phenomenal nature and mainly of their 'signal-function' are, for the time being, subjective elements of secondary importance. By virtue of having an objective influence, the phenomenal sphere secures the ontological frame of phenomenal existence. And this influence is the fundamental connecting link in man's social existence: it is his activity, his praxis. This is the way in which 'being for others' joins praxis on the one hand, and the more subtle spheres of subjectivity, consciousness and the ideological forms of perception on the other. But as a connecting link, it is an ontological datum (fact).

'Being for others' is, of course, by no means homogeneous as an active force and as a phenomenon. A transaction on the stock exchange causing a passing stir in the money market immediately brings a multiplying effect in its train, but nearly every bank, every enterprise and speculator will react in different ways. In the instance of other phenomena, e.g. regarding the case of rising wages, the 'effect for the other' is a more homogeneous medium, forcing modern capital to respond to the content of this phenomenon through the introduction of automation, by means of more intensive application of the feats of technological revolution. 'Being for others' means therefore only that the influence emanating from a particular partial totality of society, or perhaps from some of its individual members, as a phenomenon, at once functions with the force of continued effect for relationships of different content, and so enters the new environment as an independent factor, there to call forth either homogeneous or heterogeneous responses. The objectivity of the effect will be the conclusive proof of the ontological nature of appearance (phenomena), for what is decisive here is an ontological relation that can also be recognized subsequently, and not an epistemological relation within which the appearance-category is formed as the speculative and conceptual husk of an essentially different motion of reality.

This configuration is characterized, first of all, by indifference. In contradistinction to the epistemological appearance-concept, it is relatively free from interpretations since the active effect is its primary frame and only with respect to the response will the conscious search for alternatives and decision-making be of importance again. Hence this indifference means that 'being for others' is not a category of direction, it does not tend towards some well-defined relationship, nor has it the intention of entering a relationship which it has created itself; its subsequent fate is for the most part a matter of indifference to it: 'it is thrown out' into the world to be received by whatever is adjusted to respond to this particular influence. Merely a subcase of this basically indifferent directedness is the abstract trend which figures in social phenomena, a trend in which certain attitudes occurring *en masse* are *a priori* made only for a certain type of recipient — without this being determined beforehand. (Thus, for example, the shopping craze reacts as a phenomenon on those who had not perhaps the slightest intention of shopping, and the phenomenon of inflation feeds upon this collective hysteria, too.) Even here the directedness of the phenomenon develops only after the event, since the individual shopping sprees do not aim at influencing the attitude of the other shoppers. They become an influence through people's perception and teleological transposal, and a motivated goal only afterwards.

In this sense, the phenomenon merely 'is', or to be more precise, it has an effect in its form of action; but it has no bearing on how this effect develops further. This depends on the interaction into which it enters spontaneously or directed by the guidelines of the social mechanism, but which it does not itself create. To the victim who is lying on the ground after having been involved in an accident it makes absolutely no difference what sort of pedagogical, sociological or psychological after-effects and deformations this 'shock-effect' might cause in the mind of an elderly woman passing by, or of a child who happened upon the scene. But at the same time, this indifference also means the 'integration' of the effect: its ontological nature can be proved by the trauma carried perhaps for life. So the mode of 'being for others' is only the abstract range of possibilities; its concrete possibilities (what may become of it, what sort of 'responses' are created under its influence) depend on the structures of the recipients it meets. It is through this quasi-indifference that the individual partial totalities are in contact with and incorporate themselves in one another, and form the process of the general movement of society.

The category of 'being for others' means at the same time that a particular phenomenon, as the result and as the effect of the total work done by the individuals who form society or of their individual activity, is not incorporated into the other segments of society in its original totality, in its original structural quality. The incorporation is a function of the effect, of the requirements of the receiving medium. Only a few of the phenomena 'cast' into the indifferent field of reality will be selected by the various recipients, groups and spheres of action so as to transform their effect into a different quality and apply it as such a transformed energy source. In other words, the given phenomenon is incorporated only in a certain respect; it is this relevance that will then facilitate the selection, independently of the common or individual origin and totality of the human activity which 'irradiates' these phenomena. 'Being for others' means only a relational existence and a qualitative selection which is different for each partial totality. Hence we must also refute the myth of the 'constant conjunction' of reality, for it becomes clear precisely in this qualitatively and selectively effective interaction that the thesis "any thing may produce any thing" [David Hume: *A Treatise of Human Nature,* Oxford 1986, p. 173] is only a fiction.[5]

It is only through a selective relation that the objective conclusion created in the interplay of deeds, contents of action and various interests can, as a phenomenon, rejoin the life of society. In other words, this means that numerous effects may even be excluded or that they may encounter 'all other things' only through thousands of indirect intermediations, or not even that.

This selective particularity of social praxis depends not only on the responsiveness of individuals and their capability for choosing, but also on the objective screening device of their position, of their socially given 'opportunities to receive'. However, even this device may render a man 'deaf' to certain effects. Only in the simplified dialectic of dogmatism could it be supposed that a towel and an elephant are ontologically related to one another.

Hence 'being for others' is evidence that social phenomena are of the nature of existence, i.e., have an ontological basis. However, their relational existence, their selectivity (which, owing likewise to the objective system of social praxis, extends from the division of labour through the class structure to the common requirements of private life) is at the same time an objectively functioning abstraction. When we stated above that in the phenomenal sphere the totality of society is transformed onto the plane of a single relation, onto a minute surface, it was also

[5] It was Hume who discovered that conjunction was identical with the creation of a conceptual relationship. This idea was then taken over by modern philosophy, gaining acceptance through invisible transposals even in dogmatically interpreted dialectical materialism. So the thesis that 'any thing may produce any thing' is, in its origin, a universalizing possibility for conceptual action, a capacity for conceptual relatedness, which has simply been ontologized by vulgar materialism.

meant for this abstraction, for indeed this transformation takes place also on the plane of phenomenal totality. In the activity of individual persons or in the objective accomplishments of the total activity of an economically closed community (taking this activity and the objective accomplishments as phenomena) the totality of the original effects and the social totality inherent in and emanating from them can take on only a 'pictured', i.e., an abstract, form of manifestation. It cannot show these totalities in their fullness. This abstract form also mediates the essential processes (the intentions of the agents, the tendency of their total work, the particular features of their character which are 'motivated' by society), but allows them to become effective only in a single, or perhaps in a few transposals. And the element of the likewise objective 'concealment' manifests itself also through this abstraction which takes place in reality. To express it metaphorically: relationships not only 'present themselves' in phenomena, but the transposal also conceals the greater part of these properties.

The dual meaning of 'being for others', its ontological basic level and its epistemological outgrowth are evident here. From the original totality, from the 'irradiating' medium of human action the objective effects that are 'for others' can reach the other medium (though in a state of transformation). This transposed effect, however, manifests itself for the reflection as an immediate and for the time being final configuration. This is the only thing which man consciously confronts, and he cannot see — in the sphere of immediacy — the other substantive relationships and phenomenal elements which are enclosed in this transformed configuration. Therefore, we consciously identify things with what they seem and how they present themselves for us. On the other hand, in the medium of practical life we also experience the effects that the conscious reaction conceals or avoids: we respond to them unknowingly. In terms of ontology this represents the basis for the already quoted phrase of Marx's, namely that 'they do not know it, but they do it'. In this difference — the discrepancy between the real active influence and its conscious epistemological 'picturing' — there manifests itself for the first time the possibility (at least the abstract possibility) of illusions. Already in this rudimentary manifestation 'being for others' is the *sine qua non* of that possibility, the secret of its objectivity, which is the nub of its philosophical relevance. (We shall return to this problem in more detail in the chapter on 'The objectivity of illusions'.)

Another feature of 'being for others' is the receptive relation of action and cognition, i.e. the selective connection incorporating (in a transformed quality) phenomena which exist 'for others' in an 'indifferent' form. What is it then which forms the screening device for the 'recipient', for the 'other'? The epistemological answer is again much too simple. It says that it is conscious deliberation, value selection, interest, requirement etc., which raise from the indifferent background of social praxis the effects to which individuals or a community must respond. It is obvious that society as a mechanism would be inconceivable without a conscious reaction of this sort. But again, the selection does not occur only through individual 'choices'. It is implicit in the ontological structure of society that certain institutions, socially evolved roles and positions make a careful selection of, and indeed perform these effects already before they are submitted to the judgment of individuals who as formative agents shape these institutions and fill these positions. Objectification means not only that the result of individual activity assumes an objective form but also that, within a certain social order, the field of social activity organized from the bottom upwards and consisting of individual decisions and courses of action also, after a shorter or longer interval, assumes an objective form like a traffic-directing Highway Code, and will furthermore influence the possibilities of acting and receiving likewise in this objectified or, in less propitious instances, fetishized form. Capitalism and modern manipulation have increased this fetishized objectification to an utmost degree, creating the illusion that any individual action, preventive measure or initiative is hopeless because it remains *a priori* isolated. But even here the situation is that, within the given institutionally objectified limits, there remains a margin for individual or group action which can reach beyond the preformed effects of institutions and roles.

The problem is real: only as a rare exception can the immediate exchange of activities figure in social reality. It is only through a series of intervening media, culminating in objective impulses which organize themselves from partial effects into a cumulative final effect widely divergent from the original intentions that individuals, with their deeds, and groups reach the point at which they will begin to exert an influence. Consider, for example, the motion of the economic machinery. In competition the decline in supply and, with it, the rapid fluctuation of average profit and consequently the decrease or increase of the capital requirement of some particular industry form an unbroken process, but only through the objective screening device of the intervening media. The declining demand for some particular merchandise is not due to the direct effect of individual actions, still less to the fact that the factory has switched over to production at only half capacity. The system of intermediaries is the screening medium which intercepts people's individual or collective active influences, filters them and passes them onto another sector of society after blending them with its own influence, which once again comprises the objectivization of the interests and actions of only one group of people. It is therefore through these media that the influence reaches the 'other' for which the objective phenomenon 'exists'. But this intermediary process also changes its original structure in that besides screening out the indifferent elements it adds some of its own. This is another feature which is characteristic of 'being for others'. Only in rare instances does it exist as an immediate connection, for in the basic formula of ontology 'being for others' is always mediated, even if in terms of epistemology (this time on the level of false consciousness) we cannot recognize or form an incorrect opinion of the colouring, transforming effects of these media. And let us add that this system of intermediaries is once again man's doing: the established, objectified media through which the effect of some phenomenon 'for others' passes also represent the 'screening effect' of human hands, intentions and decisions, as well as their power of transformation. Therefore the fundamental law of the objective existence of the phenomenal sphere 'for others' is that it is both more and less than what has been 'radiated'. More, insofar as the intervening medium adds something of its own ingredients as it enters into interaction with it and develops the effective range of possibilities that will be received as a phenomenon, as an effect at some other point of society; and less, insofar as the gesture of adding will at the same time screen out some essential elements. That is why Hegel says that appearance — this time taken in a social sense — is always more than the original essence: "Law is the Reflection of Appearance into identity with itself ... Law is this foundation itself, and appearance is the same content, but it contains more too, namely the unessential content of its immediate being. For the form-determination also, by virtue of which the Appearance as such is distinct from law, is a content, and is likewise distinct from the content of law." [Hegel: *Science of Logic, Op. cit.,* p. 132.] With the concept of form-determination Hegel makes a clear reference to the concealing and adding function of the intervening medium. But the form-determining work of the intermediation is at the same time an enrichment, and in the course of this enrichment also a concealment of the essence, for indeed in reflection it is long impossible to know what pertains to the original phenomenon and what is added or concealed by the series of transposals of the intermediation.

To sum up the foregoing: 'being for others' represents a guarantee for the objectivity of social phenomena. By virtue of the categorial relationship, phenomenal existence exceeds the subjectivistic range of the solely epistemological datum and becomes an ontological datum.

However, not merely the objectivity of social phenomena is measured against this criterion, for indeed this category helps us to understand the problem of the ontological differences between natural and social phenomena. According to our hypothesis, as we said at the beginning of this chapter, it would lead to contradictions if we recognized this reflectional determinateness of appearance and essence only as a socio-ontological category and if we excluded nature from its domain. Although we have made a thorough examination of several weighty reasons for this exclusion, the most important of which was Goethe's concept of totality and the peculiar ability of praxis to create and select the essence, on the basis of our present grounding in philosophy there are a number of counter-arguments to which we could give no satisfactory answers.

On the other hand, if in natural philosophy we applied without further ado the category of 'presentation' developed in social ontology, we would create confusion with the more developed categorial relations of a higher form of motion. Since the social category of 'presentation' is made up also of the constitutive elements of human praxis, subjectivity and consciousness (reflection), to apply this relatively high-order category of presentation to either inorganic or organic nature cannot possibly lead to correct results. Where is the solution then? In our view, the category of 'being for others' offers also the criterion of the delineation of degrees. What is that 'other' for which a primary source becomes a phenomenon and what sort of constitutive role is played by the medium which receives this 'extraneous' effect? If we try to arrange the natural and the social characteristic of 'presentation' on the basis of this category, we must proceed from what Marx stated in his *Economic and Philosophic Manuscripts* on the relation between natural phenomena and the 'other' — likewise a receiving medium of nature: "The sun is an object for the plant, an indispensable object which confirms its life, just as the plant is an object for the sun, an expression of its life-awakening power and its objective essential power." [*Op. cit.*, p. 390.] The sun's existence for the flower can be regarded only as a light-source, though this single relation also contains all of its other characteristics, transforming the totality of its influence-exerting potential on earth onto this single feature, for only in this way can it be 'life-awakening'. Here 'being for others' still asserts itself on a lower level since it is obvious that its phenomenal existence is not shaped by the subjective, acting and conscious relationship, nor by the selection of social intermediation. At the same time it can be assumed that with regard to the functioning of the sun the 'light-effect' is in part one of the forms of presentation of physical reactions, just as its life-awakening effect describes phenomenally the relation between the forest and the undergrowth in its shade. Even 'shade' enters as 'being for others' into the life of whatever is growing on the ground, albeit originally it made no difference what further effect was produced by the light and shade effect, which is incidentally a characteristic of any optical relation between light and some object. This original source nevertheless affects the quality of the undergrowth through the transposal of an objective relationship represented by the forest, the sunshine, the number of rays glimmering through, the culture of bacteria types determined by these factors etc. It becomes one of the phenomena of the relation between the sunshine, the forest and the culture. Let us suppose, without intending to prove it now, that the category of presentation is a general ontological structure, the presentational form of all homogeneous totalities (e.g. the traffic system of a town irrespective of the pedestrians, the living conditions of the drivers, the condition of the streets, the condition of war or peace etc.) as well as of heterogeneous ones as they 'exist in one another' — that is, the presentational form of all relational connections. In other words, since things are connected with one another in complex ways and since it is an ontological impossibility to separate them distinctly, the mode of existence of this correlated existence is none else but presentation, the effect that 'is there for others' and can act further, and which to a lesser or higher degree differs from the structural connection which guides this mutual existence in this single relation. At the same time this supposition entails also that the form of presentation, phenomenal existence, displays a different configuration, a different structural relationship in each form of motion. The most problematic aspect of the form of presentation is to be found in its role in inorganic nature. (For instance, where, in the case of a chemical reaction, do the sphere of appearance and the sphere of essence separate, as can be observed in society in the case of immediacy and general movement? And, in general, every chemical reaction is merely a 'directional phenomenon', the description of a selected relation according to natural laws, whereas in nature these relations take effect simultaneously, and it is precisely these 'simultaneous forms of presentation' which chemistry as a science cannot grasp.) But even in the case of other forms of motion of inorganic nature, such as mechanics and thermodynamics, it is extremely difficult to grasp a configuration of appearance which might tell us more than what the simple dialectical categories of 'cause and effect', 'antecedent and consequence' or 'interaction' describe. And while in our efforts to reconstruct Marxism we also object to the impoverishment of the philosophical categories, we would surely simplify the system of our concepts if we were willing to reduce everything to a single basic relation or to reduplicate the same categorial relationship. (The relation between cause and effect may also mean the relation between appearance and essence.) For all that, there are also arguments in support of the possibility that this category may win an objective acceptance in an elementary form, measured here, too, by the criterion of 'being for others'. (For example, oxidation as a substantive process, as well as the 'being for others' of the heat effect bringing about melting and evaporation in surrounding objects.)

Therefore, if we do not want this category to relapse into the epistemological mode of understanding, we must postulate the primitive forms of its presence even in inorganic nature. The crucial point is the historical gradation which prevails between the individual forms of motion and which also changes the configuration of presentation and its categorial functioning. We can see the elementary degrees in the respective developments of organic nature and human society. With the manifestation of elementary consciousness — in highly developed organic nature — there enters the subjective component of cognition and even the dialectic of society-linked appearance-creation and essence-formation. In the light of this series of historical developments we are of the opinion that we can reject the antinomy mentioned in the Introduction, which claims to discover ontological contradictions between nature and the mode of presentation in society. The mode of presentation operative in society, with an epistemological tinge but essentially of an ontological character, could arise only through having already existed in the previous forms of motion embryonically, in elementary forms of 'being for others'. This provided the foundation for and was transformed by the coming into play of consciousness. If we do not take this difference in degree into consideration, i.e., if we disregard Marx's methodological instruction according to which historicity also applies to nature, we will be inclined to exclude the presentation-category of nature. On the other hand, if we include this historical phase, we can eliminate the simplifying and essentially dogmatic notion that the law of presentation prevails in the same configuration in the various forms of motion and that just as in human society so in nature there is an appearance-category functioning which is hypostatized by a subject.

In modern bourgeois philosophy we are witness to a contrary process. The appearance-concept that has developed here (as a matter of fact, through the revival of the influence of Kant) has made the natural philosophical, hence the epistemological type of presentation a model for society. Here, too, the distinction between the different forms of motion is blurred, but now merely the reflexive relation of the knowing subject to nature becomes the constituent element of the phenomenal sphere. The modern concept of 'phenomenon', as we shall see in the next chapter, is intended to describe society and nature together, solely in the configurations of their presentation. At first sight it looks like another name, a synonym, for the appearance-category; actually, because of its undifferentiated character which confuses the two basic forms of motion, it may become the very means by which the appearance-category is befogged. So the 'phenomenon'-concept is rather problematic, not only because it is an attempt at objectivizing immediacy, sheer presentation, and thus also illusory relationships, but also because eventually it obscures the characteristic feature of social presentation, helping to conceal its illusory forms of presentation and to paralyze the endeavours to disclose these illusions. The overlapping of these two ranges of validity did not begin with modern philosophy but during the period of the British Enlightenment, in the theory of primary and secondary qualities; but after the disintegration of Hegelian philosophy it was formulated more emphatically and directed against social ontology. It is therefore typical that F. H. Bradley, who was schooled on Hegelian ideas, and was the first to recognize illusion as a philosophical and social problem, salved his conscience with this (Lockean) solution in terms of natural philosophy. [Cf. F. H. Bradley: *Appearance and Reality,* London 1906, p. 11 ff.; p. 184 ff.] But the stripping away of this categorial model, of this conceptual bias, is rendered possible only by on the one hand proceeding from the differentiated development of the appearance-problem, an increasingly complex development which varies with each form of motion, and on the other hand successfully isolating and analyzing in their independent nature the genesis of false appearances (illusions) and their laws. Just as there is a connection between prolegomena and the essential question, so there is one between the more general question of presentation, which goes far beyond the scope of this study, and our specific subject: illusion.

1.2. THE SEMANTIC RANGE OF THE 'PHENOMENON'-CATEGORY: THE DIALECTIC OF 'BEING FOR US'

Before proceeding further we must concern ourselves with a critique of the concept of 'phenomenon', the best-known appearance-category in philosophy. This approach is justified both by the nature of certain modern tendencies in philosophy and by the logic of our foregoing discussion. In our view, the appearance-category corresponds to an objective ontological concept, primarily in a sociological sense, but also, on an elementary level, in certain natural relationships. The 'phenomenon'-category, on the contrary, by virtue of both its origin and its function, is epistemological in character. It records and renders objective for the consciousness of an age the various phases of 'being for us'. Phenomenology and, then, existentialism — faithful to the attempt started with Kant — made this epistemological character indistinct, endowing it with an objective, ontological role. So the difference has disappeared between the natural and the social forms of motion on the one hand, and between the relations of conscious and ontological appearance on the other. Hence it follows that we are confronted here with a double task. We must, on the one hand, mark out the limits of the true meaning of the 'phenomenon'-category, the range within which it can function with an objective reason, and, on the other, reveal this concept's basic affinity with natural philosophy and, in connection with this, its epistemological quality, and thus strip away from it the ontologizing deposit with which it has become encrusted in the course of the history of philosophy, the disguise on which modern thinking has conferred a certain permanence. In the course of this criticism our main objective is therefore not to prove the emptyness of this concept, which would be an absurd aim, but rather to elucidate the difference between the truth-contents of the two concepts, appearance and 'phenomenon', while devoting some attention to the exchange of concepts in the history of philosophy.

The first stage of the erroneous identification of the appearance-concept with the 'phenomenon'-category can be queried with relative ease. The only thing that deserves attention here is that a similar process has occurred in both the more recent trends of bourgeois philosophical thinking and the dogmatic interpretations: there has emerged the biassed view according to which in the course of cognition consciousness first encounters phenomena and only then proceeds towards the revelation of essence. Thus it directly confronts the phenomenal sphere, or, in other words, a mass of phenomena. But was the rock, which, lying among other pieces of stone, turned up in a man's path, and which, in the light of our present knowledge, 'refers' to or indicates the oil deposits hidden in the depth of the earth — a phenomenon? In terms of ontology it was. Epistemologically, and indeed from the point of view of the metabolic exchange between man and nature, at that time it was still only one thing in the colourless and inconspicuous medium of all other — indifferent — data forming an objective background. Even when man stumbled against it, he struck his toe against a piece of stone that to him looked the same as the other ones lying beside it; it held no 'signal function' for him as yet. At the same time, this view was combined with a Platonizing conception of harmony according to which the gradations of the cognitive process (perception, observation, abstract thought) corresponded to those of reality and its arrangement in depth (the path extending from appearance to essence). However, and here at once there arises one of this book's fundamental problems, reality is not layered like an onion, and the common experience according to which the progressing act of knowledge is delving 'inwards' in matter cannot be raised to the rank of ontology.

On the other hand, this proposition is not true in terms of epistemology either. Man encounters first the undivided mass of things rather than phenomena. These data become phenomena only through practical work, when, by the aid of practice, science and technological progress, a systematic division is achieved in which we can discern what is 'behind' the thing immediately in front of us, i.e. essence. We speak of phenomena only when the cognitive process has raised from the indifferent material background — at least in segments — the order of the intrinsic relations, an order which connects the various things, configurations and motions. The 'thing', 'object' or the 'event' is still not a phenomenon. They become phenomena by virtue of the ability of praxis and cognition to create a relation, though in its objective, ontological structure any 'thing' can possibly be a part, and so also an 'appearance', of a network of numerous relationships.

At the beginning of the cognitive process there is only the still undivided category of 'phenomenon', but this configuration is not identical with appearance taken in the Hegelian sense. The Hegelian concept is objective, whereas 'phenomenon' implies the 'husk' of subjectivity, of an activity consequential upon praxis, of the sometimes arbitrarily established relation, of the objective connection and ontological form between things, objects and occurrences, as well as of the subjectivistic hypothesizing work of the first steps of praxis. The category of praxis is a configuration in which subjective and objective elements are intermixed.[6] Ample examples of this sort are provided by the history of philosophy. Let us once again refer to the case of the picture men formed of stars and planets in the course of investigations carried out by means of primitive — astigmatic — telescopes. These first optical instruments produced strong distortions. The astigmatic lenses drew concentric circles around each light source, whereupon there were 'rings' perceptible round each star. Yet at that time the astronomers still did not and could not know about the distortion, and so they believed that what they saw, including the phenomenon-distorting effect of the instrument, was the properties of the stars. They thought, because that was what they saw, that every star and planet had a ring. Only later on, with the discovery of the anastigmatic lenses, did they eliminate this error. Then they realized that this peculiar 'phenomenon' had been caused by the lenses and 'projected' on to the stars and planets, and that here the subjective, or more exactly, the empirical approach (instrument and observation) had become entangled with the objective phenomenon, the light effect of the stars. (And it was also found at this time that only Jupiter had rings, not because of the distorting effect of the instruments but owing to its characteristic structural composition.) Here the 'phenomenon', identified in the course of the cognitive process with the phenomenon, resulted from the still undifferentiated unity of subject and object, from the interaction between practical consciousness applying various devices (the historically and technically given standard of cognition) and the objective phenomenal existence, an interaction which cognition objectivized and attributed to objects themselves. This form is then the elementary level of subject and object as we stated above and to which Lukács referred in his *History and Class Consciousness.* This unity may materialize on various levels, whereas on the elementary level it may give rise to misleading configurations pregnant with illusions. The reason why the 'phenomenon'-category needs criticizing is because the modern use of concepts has taken no account of the unity of subject and object materializing through these stages and ontologized the immediate

[6] Hence this first, 'mixed' configuration of praxis and cognition — including both the effect of the existence of the subject and the complex interaction of the objective phenomenal sphere which it creates — is what in this work we propose to call a 'phenomenon', while perfectly realizing that in both classical and modern philosophy this term has other meanings and refers to different relationships. We shall return to these differences.

(abstract) unity here. With this, however, it smuggled subjective elements, too, into the appearance-concept.

The example illustrates some very important primary characteristics of the 'phenomenon'-category.

First of all, this configuration is nothing else but the way in which an objective phenomenon (a stellar constellation) manifests itself. It is not a 'mere' illusion. It is rather the result of the relational effect of an existing objective datum (fact). But this effect is distorted by the mediation of practice and cognition, whilst at the same time it represents this distorted form of manifestation 'for us' as the property of the object. Accordingly, the element of apparency is already present in the 'phenomenon'. And it is also evident that this illusion is the offspring of empirical relations, that the picture beginning to take shape is a function of the instruments provided by scientific and technological achievements, a function of the level of development attained by human practice. Furthermore, it is also obvious that the false form of manifestation seems to be the property of the object (star) because we have at our disposal only this single mode of approach, and we cannot check the information obtained by any other means. Thence it follows naturally that with the progress of science, with the perfection of the means of research, or through the invention of new ones approaching the given object from another angle, this distortion would be eliminated, and so the element of apparency which is present in the 'phenomenon' would also disappear. (We no longer believe that there are rings round every star.) Thirdly, it is also clear that the 'subjective side', i.e., the working process of 'cognition', comprises not only consciousness proper but also the entire apparatus of practice, the historically available means of cognition, and that the subject faces reality in this more complete state as a historically prevailing subject-being. So it is on the basis of ontology that it is possible to understand the process of cognitive progress as well as the origin and dissolution of illusions, i.e. our constantly changing epistemological relatedness to the world. Finally, the 'phenomenon'-picture formed in this way, the conception passing into science or common knowledge and controlling praxis, also acquires a certain type of objectivity. It is not a simple error, but since on the given level of cognition the world only looks like that, these notions become identical with reality and are incorporated in the process of the metabolic exchange with nature. They are improved or stabilized, but for a longer or shorter time they function unchanged and their performance in this role still further confirms the credibility of their 'objective' verisimilitude. If that be the case, 'phenomenon' is not a subjective idea, illusion or perhaps a copy of reality consciously regarded as being ambiguous, but it is the authentic picture of the 'one and only' reality — although in terms of ontology it is only one stage in the history of the praxis of mankind and in the related cognitive process, and includes both the true structure of the given relationship, object or phenomenon and the 'picture' which is formed of this structure and subjectivized to the extent that it is full of distortions arising from the possibilities for the acquisition of knowledge. This duality is the first contradiction of the 'phenomenon'-category. It simultaneously comprises the subjective marks of both the real objects and the praxis level of the historically given cognitive capacity.

However, the case illustrated (the development of astronomy) is still of a specific character, and it cannot render suitably perceptible the structure that is characteristic in general of the cognition of the 'phenomenon'-category. We try to choose our subsequent examples so that light be shed on the more general characteristics as well.

We would include in the first 'phenomenon'-type those forms in which it is usual to comprise the decisive role of the 'first mode of manifestation' in defining the limits of phenomena. It is a matter of common knowledge that in the course of a new experiment we are inclined to regard all the conditions of the 'first success' obtaining at that time and its collateral facts as the necessary appurtenances of success itself. When we are trying to open a lock or repair a defective clock we think that 'we should go about it in the same way as we did on a previous occasion'. And indeed, we do try to reconstruct all conditions of that time. In other words, we identify the random modes of manifestation of things, the incidental and practice-formed relationships in which they manifested themselves, with the things themselves, with their mode of existence. It is obvious that a good many of these conditions pertain to the success, but there will also be among them a number of incidental, even negligible circumstances that we nevertheless want to retain. The 'phenomenon' makes these two series of conditions one and the same. By virtue of the fact that it cannot as yet separate the random conditions from the necessary ones it regards both as the concomitant circumstances of the things. The order of relationship of the first approach is abstract because it is interwoven with numerous contingent elements, and it is only slowly reduced to the point of recognizing the concrete appurtenances of the true mode of manifestation. This process also takes place within the unity of subject and object in praxis. The more aspects of an object we try to grasp, the more obvious its 'noumenal nature' and the probability of its being related to the subject will be. Thus it loses its outer crust.

This 'phenomenon'-type is characteristic of magical thinking. The traditionally preserved forms of cultic rites can be accounted for by the fact that people are bound to cling to the formula which has once already resulted in success, and so they are apprehensive about departing from even the slightest details. The reason for this apprehension is of course easy to understand, namely that they have happened to light upon the true relationship which led them to a useful result and must therefore firmly retain every contingent circumstance. Objectively they cannot know which one of the elements produced the favourable result. The numerous contingent cultic and mystical gestures not only revealed the given and useful relationship, but also concealed it by means of their contingency and their (rationally) puzzling nature.

Such an attitude is also discernible in the simplest technological achievements. Gordon Childe, in his *The Cradle of Civilisation* describes e.g. how the first potters and iron-smiths kept secret numerous methods which had no concrete role in the process of technology but which they had apparently met in the course of their first successful attempt, and so they dared not forgo them. He writes that no single, general potter's craft existed then, and that there were as many traditional tricks of the trade as social groups. Even if it seems to us that these were variations of a theme, the women handing them down could not differentiate between the necessary activities and the random modifications. The technical specifications of the barbaric science were, in all likelihood, accompanied by innumerable unnecessary magics and ceremonies. The same structure of practice and consciousness can be observed in alchemy, in the initial stages of the medical sciences or of the study of electricity.

To what extent does this configuration have the character of a 'phenomenon'? To the extent that the material being formed in the working process seemed to have precisely the kind of structure that the moulding hand allowed to be seen in the first successful solution. Its 'phenomenon'-form is constituted by genuine relationships as well, but many contingent elements related to the exploratory, seemingly random steps of empirical discovery have also stuck to this structure. The conditions of the first success arose from the process in which the subjective side, i.e. the contingent elements of practice and the unknown effects of the means employed, interlaced with the concrete forms of the things' behaviour.[7]

We might designate the next 'phenomenon'-type as the screening effect of the means employed. Countless versions of this type can be seen in the historical development of the sciences. Some new levels of reality are made accessible through the applications of new means of investigation (microscope, cloud-chamber, X-ray etc.). This means at the same time that until the microscope was invented the idea of the cellular structure of tissues could not even arise and people could not have even the slightest idea of the existence of microbes and viruses. Hence the historically given technological level of the instruments also determines the conception that science has of reality and of the world picture with which it identifies this reality. Science develops between objective limits. It can see an ever increasing proportion of the totality of true relationships, as much as is permitted by the given standard of the instruments of investigation. With the invention of the microscope a whole new level of reality was revealed: some earlier views based on 'phenomena' have vanished, others have been enriched with new relationships, and cognition has managed to registrate numerous objective phenomena. But even the new instruments set a certain limit to the scientists' cognizance, and so reality has once again manifested itself only through a 'gap', though a wider one at this time. In the overall picture we have managed to form on the basis of this manifestation there are again 'phenomenon'-like configurations. The means employed once again screen out a number of relationships, they often afford us a glimpse of reality in only one of its aspects, only one of its mode of manifestation, and consequently our ideas that are based on it also remain biassed. But this bias is not an individual error. It is an objective bias, for it is not through an arbitrary decision that we define the field of view of the available instruments but this is, after all, what the world objectively looks like through this particular 'system of lenses'.

[7] The phenomenological method of investigation is characterized by this 'phenomenon'-type, by the fetishization of the 'first manifestation'. Adorno wittily points out how in Husserl's case this first, immediate datum is identified with the total phenomenon: "The first of the philosophers makes a total claim: It is unmediated and immediate. In order to satisfy their own concept, mediations would always just be accounted for as practically addenda to thought and peeled off the first which is irreducible in itself." [Th. W. Adorno: *Against Epistemology: A Critique,* transl. W. Domingo, Oxford 1982, p. 7.] But, says Adorno, factuality can never be such a 'first' (immediate) datum, for every material existence is mediated. Whatever claim it makes to be scientific, its 'phenomenalistic' nature, if it builds only on this abstract 'first manifestation', will be pregnant with illusions.

Here we can see that though the 'phenomenon', as regards its origin, is the offspring of praxis and a subjective category, yet it will become an objective configuration in its function. The real phenomenon and the objective possibility of an empirical and conceptual approach jointly show us the immediate object of cognition, that is, the 'phenomenon'. Only when science reveals some other aspects of this 'phenomenon' from a different angle, i.e., through different means, will it then develop from the multifaceted nature of these interactions what pertains to its phenomenal nature and what was caused by the distortion and the screening effect of the instruments. These 'phenomenon'-configurations manifesting themselves in the world of science are not only the errors of the historical past. One of the main subjects of the scientific and philosophical arguments is concerned with the reciprocal effect between micro-instruments and micro-phenomena. On one side there are those who adhere to the irresolvability of this reciprocal effect, and on the other side there are efforts to develop a new method of research (for example, statistical atomic theory) by which one might avoid the effect which the means of research has on the phenomenon.

The list of deficiencies of the 'phenomenon' as a conscious, 'known' picture of reality is augmented by the concomitant effects of the means of research, effects which have been left out of consideration and which we identify with the phenomena. A classic case in point is the primitive theory of free fall. It was not only Aristotle who thought that the velocity of the fall of a body depended on its weight; this error of observation had, subsequently, been repeated over and over again. But behind the error there lies a 'phenomenon'-like appearance: man investigated only the relation between the bodies and the ground, and he did not take into account the resistance of the air. A mistake of a similar type was made by Helmont, who is known as the first to have arranged the various gases in a system, but who simply did not regard air as a gaseous matter. It was a natural medium, so it remained 'invisible'. János Selye sees things in their true colour when in connection with the analysis of a series of experimental errors he states that the difference between the active ingredient figuring as a cause and regarded as a cause is what leads to false results.[8]

[8] In his *From Dream to Discovery: On Being a Scientist* [New York—Toronto—London 1964] he writes that experimental medical science is based on the principle that if a certain change occurs only in an individual case in which a certain treatment has been given, then that treatment is to be regarded as the cause of the change. But it could be the source of an enormous error if the effect had not necessarily been caused by what we had administered.

He furthermore points out that the coextensive and, thus, ignored factor is represented by the deficiency brought about by an active ingredient or a deficiency in general, while the experimenter thinks in terms only of active and not negative effects [Selye: *Op. cit.*].

We can see everywhere that cognition wants to work only with abstract objects, and that it does not notice and by reason of its teleological constraints cannot even take account of the accompanying and for the time being still invisible active constituents, the properties of which it regards only as the characteristics of the already discovered medium. Guided by practice, cognition naturally transcends this limitation. If it is successful in approaching this range of phenomena from a different relationship, too, then the 'concomitant' phenomenon and the original 'conveying' medium will at once be separated from one another. And with this, illusions will also disappear.

These many different kinds of 'phenomena'-types remind us that this category encompasses a wide range of the degrees of objectivity. In all these cases, we have been dealing with the forms of being for us produced by the effect on reality of our 'subject-being', an effect in which the subjective and the objective elements are intertwined. But at one pole the objectively phenomenon-like predominates. (This class includes the archetype, the 'phenomenon' of the star or planet seen with rings around it on account of the astigmatic lenses.) At the other pole the subjective element is given a predominant role, either by the instruments of 'subject-being' 'producing' the greater part of the given 'phenomenon' or by cognitive consciousness resorting to its analogical reasoning to 'explain' the phenomena that have already undergone preparation. (An example of this is the development of the mechanical world picture.)

As opposed to the 'phenomenon'-category which has developed in the history of philosophy, ours is the immediate reflection of the practical unity of subject and object and forming at the same time the scope of the progress of cognition. This consideration rests on Lenin's idea that the relation between consciousness and reality is absolute only in terms of epistemology and that the rest of their relationships are relative: "Of course, even the antithesis of matter and mind has absolute significance only within the bounds of a very limited field — in this case exclusively within the limits of the fundamental epistemological problem of what is to be regarded as primary and what as secondary. Beyond these bounds the relative character of this antithesis is indubitable." [V. I. Lenin: 'Materialism and Empirio-Criticism', in: *Collected Works,* transl. A. Fineberg, Moscow 1962, Vol. 14, p. 147.] We have tried to emphasize the primacy of the material world precisely by showing the 'superior force' of reality in human praxis: the reflection wants to see clearly in the matter of new phenomena 'breaking in' time and time again. But the mode of manifestation of reality is already determined by the manner in which practice makes its approach, and in this medium the difference between the subjective and objective sides becomes relative, thus creating the relative objectivity of the 'phenomenon'. It seems to us that what has been transferred on to the objects and processes merely by practice, or by its instruments, historical conditions and collateral facts, were the characteristics of the objects themselves. However the subsequent progress of cognition and practice will clear away these transferred elements, but only in order for the entire process — the interaction between the epistemological and the ontological spheres — to be repeated on the higher historical level of praxis.

What has been said so far has been an approach, largely by means of examples, to definitions of the increasingly blurred character and consequently to the contradictory contents of a particular category. Let us now try to give a summary of these contradictions. First of all, the 'phenomenon' is not a phenomenon, it has none of the latter's objective relationality. It is not a 'reflexive determinateness' such as Hegel shows the category of appearance-essence to be. In other words, it has no 'counterpart' because it is born of the interaction between praxis (and with it the historically given level of subject-being) and the true material and human world. It is thus the 'phenomenon'-category which comprises that unity of subject and object which Hegel's system imposed on reality, only in a distorted shape, as an ontological system. This 'short-circuited' relationship was dissolved by Marx's theory of practice and reality, in which the practice of the historically determined 'human species' is conceived of as the 'social reality as it is in itself', actively and objectively. In our view, in Marxist thinking praxis repeatedly forms its own system which is historically 'closed' and at the same time capable of being 'exploded', i.e., open to the future, a system in which subject and object become 'identical' and the 'phenomenon'-concept is the unqualified and unconditional synthesis of this constantly developing, progressive advance which articulates this portrayal of reality and enriches mankind. The uncertainty of the history of philosophy with regard to this concept stems from the fact — to which we shall return in a separate chapter — that it could not reconcile this duality, the practical unity of subject and object and its constant 'explosion', with one another and so it hesitated between two sham solutions. By reference to the momentarily existing apparency (illusoriness) enclosed in the 'phenomenon'-concept, it identified this category — and worse still, the world picture based on it — with illusion itself, or, relying on the likewise momentary contentual elements of being for us *(für uns Sein)*, it raised it to the rank of an objective appearance-concept. Neither the former nor the latter solution could possibly bring the issue to a successful conclusion because both conceptual versions were split apart by the moving, changing and developing nature of this category and by its tension which combined within itself reality and the subjectivistic factors deriving from praxis and postulating activity.

But on what grounds can we revive the conception of the Hegelian 'unity of subject and object' if we once aim at setting on ontological bases both the theory of appearance and the 'quasi-real' existence of illusions: their formulae in social life and in false consciousness? In line with our conjecture, the first steps of human praxis as well as all extensions of social activity and the metabolic exchange with nature occur within the 'husk' of this relative unity. The 'phenomenon'-category is the synthesis of the movement which consists in our coming into contact first in the process of cognition not with 'noumenal' relations or mere objects but, in the course of the creative work of praxis, with the as yet undifferentiated unity of object and subject. Only after reflection and the development of a multifaceted connection with things will the subject break away more explicitly from objects, separating the essential from the unessential, and be capable, from the point of view of epistemology, of finding his way about in the dialectic of appearance and essence, and in general, of using the definition of this relation. We must note here, too, that this 'initial' phase of subject and object, recurring time and again in human praxis taken in a historical sense, does not take on the same character in our relation to natural phenomena as it does in relation to social phenomena. The common ontological basis is the objective nature of reality. But concerning the 'separation' of natural phenomena from the partially erroneous and partially true images of subjective ideas, and even more so in the interrelationship of the natural processes independent of human contribution (as, for example, in the interaction of two compounds combining in a chemical reaction, a 'being for others') the essence-category is much more segmentary and relative than in social praxis. But the natural objects, existing in the midst of thousands of relationships, perform at the same time functions of both essence and appearance, and in terms of categories the unequivocal separation of essence and appearance is also problematic. The appearance-category here — now we are stating our conjectures and our largely unsettled scientific doubts —

has the characteristics of epistemology rather than those of ontology. Goethe's already quoted famous maxim, *Natur hat weder Kern noch Schale* (nature has neither kernel nor shell), gives a fairly realistic and plausible description of this relationship. The natural appearance-category has a different ontological structure since the form of motion in which it takes effect is different, too. As this range of problems is no more than an outlying branch of our discussion and one which has led to no satisfactory solution, let me refer now only to a single albeit fundamental difference. Those forms of presentation which manifest themselves in the life of society require not only a consciousness which registers, understands or misunderstands them, i.e. the subject, but also the active 'responsive man' who does more than merely record what presents itself, he understands it as a phenomenon (and, in this respect, also ontologically realizes it). In nature, on the other hand, 'being for others' *(Sein für Andere)* is likewise a basic category of ontology (thus, e.g., the 'broken-down' forms of this category are also shown in the dynamics of interaction, relation, effect etc.), but this 'being for others' means the immediacy of the effect and its homogeneity rather than the manifestation of something for consciousness, i.e. understanding and interpretation or misunderstanding. It means that by this the immediacy also gains a different meaning, which is just as matter-of-course as the fact that the structure of the phenomenal also assumes a different (inferior?) aspect. To which is added — secondly — the special feature of social phenomena. Through consciousness and action they assume the role of objective representation. For indeed, the separation of phenomena is also the result of man's practical penetration, of the selection of a particular system of segments in which the relationship between appearance and essence has an objective ontological meaning only as some relational category of direction. The phenomenon-relation taken absolutely separately, when it is not directed to man and society, cannot describe nature adequately. On the other hand, this is precisely the contradiction described by the 'phenomenon'. It determines the steadily increasing process and 'essences' of man's metabolic exchange with nature as the unity of the true ontological relationships, grasped through the elements of error, misunderstanding and apparency, with the subjectivistic 'externalizations' arising from subject-being, i.e., as the unity of subject and object in practical consciousness. In this respect, the 'phenomenon' falls within the domain of natural philosophy, it is its 'appearance-category', if we may be permitted to make this crude simplification. This is the categorial form in which the expansion of praxis captures nature's various forms of motion and builds them into a system of the total activity and the conscious accomplishments of the human race. This origin in natural philosophy can be best understood if we consider the life expectancy of the scientific 'viability' of the 'phenomenon'-formulae. Recent discoveries almost immediately modify the phenomenon-picture grasped as a 'phenomenon' and teeming with subjectivistic elements. They are inclined towards stripping off from the 'phenomenon'-picture developed in this manner anything which is a historically contingent 'externalization' of the subject-being, necessary at one time in terms of the totality of society. No such quick switch-over — such a fast-paced clarification of the phenomenon-structure — is conceivable in the domain of social practice and social consciousness. A new discovery made in sociology cannot right away put an end either to the ideological system of prejudices which have been formed until then or to the validity of earlier laws which prove to be erroneous now. Moreover, the acceptance by science of a particular new theorem can, for centuries on end, coexist with appearances and illusions of false consciousness which continue to thrive and proliferate in everyday life. Since in a society it is people who make up both the objects and the subjects of activity, the 'initial' unity of subject and object also assumes a different aspect. Here the process of becoming essential, the objective essence taken in terms of ontology, is a layer that is more tangible (although more difficult to explore), because it is a 'homogeneous' medium rather than merely a single segment, thrown into relief by the requirements of praxis, of a nature which is indifferent as compared to the life of society. Here the 'phenomenon'-category makes room for the active relationship of the dialectic of essence and appearance: for human selection, for the functioning of essence-related processes coming from the deeds of people but having command of them nevertheless, and for the immediate forms in which these processes present themselves for the individual and which are distorted in immediacy. As a concept of natural philosophy the 'phenomenon'-category can no longer comprehend these dialectics originated in a superior form of motion.

That is presumably the reason why phenomenology could obtain (mainly methodological) results as a discipline having an influence over the natural sciences, although a sociological elaboration of it proved next to impossible. The expansibility of this category has been hindered by the contradiction implied in the 'phenomenon'-category, in this case by the ontological difference in level between natural and social existence. When the sociologization of this concept nevertheless takes place, as recently with Sartre, it either leads to questionable results or necessitates the abandonment of the category.

The second contradiction involved in the 'phenomenon'-category is that notwithstanding its epistemological origin, premised by the historical data of cognition, it manifests itself as an ontological formula. Ontologization, however, follows not only from the ideological aspirations of modern philosophy, but also from the 'self-movement' of this category. The 'phenomenon', as we have seen, determines the 'for us' *(für uns)* fragment of reality, that particular condition which is shown for our consciousness and our cognitive power by the interaction of our subject-being and the objective data of nature. Regarded ontologically, and here we base our analysis on György Lukács's investigations, the 'phenomenon'-category has an indifferent attitude towards existence. It may include subjective and objective elements alike, and it may represent the various levels of this mixed configuration.[9] The reason why the 'phenomenon'-category could be in part the modern, theoretical starting-point for the absolutization of false appearances and in part a theoretical obstacle to the ontological elaboration of presentation was that the dialectic of the identity and disparity of these two spheres was not carried through. Bourgeois philosophy, fascinated by the rapid development of science, failed to take this step. Marxist thinking, on the other hand, half-heartedly balked at the difficulties involved in reconstructing the contexts of Marx's indications and, relying on some of Engels's ideas which were worth pursuing philosophically, considered the problem to have been settled once and for all. With this, however, Marxist ideology laid down its arms. In part, it invalidated its arguments in the face of the aggressive, and sometimes objectively well-aimed onslaught of phenomenology, and in part it abandoned the struggle that the differentiated and autonomous elaboration of the Marxist theory of presentation would have required.

Reverting to our problem: in the last analysis the 'phenomena' form merely *'für uns'* categories, except that in some historically 'private'

[9] Nicolai Hartmann explains this 'indifference' by saying that "it pertains to the essence of the 'phenomenon' that it has an ascertainable factual character, but the factuality of what constitutes its content cannot be ascertained." [Verbatim translation. Hartmann: *Zur Grundlegung der Ontologie,* Berlin 1965, p. 152.] We would further refine this formulation as follows: 'phenomena' represent a particular order of objectification that is indifferent to the nature of existence. It can be saturated with elements of existence and blended with fictions, as for example in the case of objective social appearances (illusions). It is this ontological difference which separates them from social phenomena which have an ontological existence and an objective effect. Only in the process of reflection can 'phenomena' occur.

aspect this 'being for us', whose ontological structure and existential configuration indicate a different formula, takes on a certain 'noumenal character'. What, with regard to its essence, is an epistemological configuration dons an ontological 'noumenal' mask at the given moment of praxis. But this mask is, once again, not merely a sheer illusion, it is built out of the elements of reality. First of all, man exists in the world not only as a knowing subject, but also as an active, flesh-and-blood individual, and the object of his activity is not the abstract natural object but (if we confine our attention at this point to the interpretation of the 'phenomenon'-concept in terms of natural philosophy) the objectivity which he forms through the means he employs, his historically 'allowed' level of these means, and his social wants, an objectivity to which even nature itself can respond only in a historically limited way. Only this 'responsive' reality which has already been formed through social praxis exists for cognition. But on the level of the immediacy of praxis, the man of practice cannot decide whether a given phenomenon is the manifestation of the objective structure of reality or merely the form-giving and challenging effect of his own activity, of his mode of approach and the means he employs. The responses are objective in both cases. They are not the products of imagination, so for the time being we cannot separate what belongs to the perplexing effect of our subject-being (of our instruments, our ability to produce an impact on something etc.), and what constitutes the true mode of manifestation of things. Hence the 'phenomenon' is not simply a false appearance or an empty, deceptive notion; on the contrary, it is a form of 'being for us', a form which at the same time embodies existence of our practice for reality and determines this relation as a 'noumenon', as an ontological fact.

But the reason why for a shorter or longer interval the 'phenomenon' can take on such an ontological 'mask' is because in it the natural relationship which 'exists for us' through our concepts (we know it and use it) is truly an objective configuration, though enclosed in subjective elements. The 'geometrical projection' of a particular phenomenon which it turns towards us may conceal perhaps its other relations, thus presenting an illusory picture, but the element which is facing us existed in the same way as the totality of the given objective connection did. If at all, then here it would be necessary to introduce certain 'degrees of being', for indeed the stars which 'present themselves for us' through the astigmatic lenses do 'exist', but not in the particular form of presentation which was then registered. But in the sense of the elementary degree of being, reality for us pertains to the category of being (and not to the range of fantasy or of the subjective realm of imagination) just as the 'noumenality' of the relationships which have been methodically explored. Hartmann wittily demonstrates that 'being for us' does not differ ontologically from 'being in itself': both exist, and perhaps only the cognitive relation will make a distinction insofar as 'being for us' is not only the geometrical projection of objects which is objectively turned towards us but also includes the distorting effect of the cognitive apparatus. Hartmann is, however, wrong in at the same time absolutizing the ontological equality of the two different types of modes of being.[10] But in my opinion it would not be inconsistent with the spirit of Marxist ontology to premise degrees of being which would likewise include noumenality, though indeed on a different level of value. The way in which *Capital* is set out proceeds from the commodity to the analysis of production capital, and then, with the introduction of monetary capital (credit, trade, transport), to the analysis of the processes of the total society where an ever newer, ever increasing range of totalities is included in the analysis, allowing ever new 'degrees of being' to find expression, thus further developing the brilliant conception which was evolved from the developmental level of Hegel's concept of reality as *Sein, Existenz, Realität* and *Wirklichkeit,* a conception which with Hegel was still rigid as he simplified it to an idealist gradation. As an element the 'phenomenon' has a 'degree of noumenality' not only in its historically inevitable configuration of objectivity but also in its form of presentation 'for us'. This degree is then absolutized, developed into a system or put to practical use by the prevailing commonplace as well as by the scientific way of thinking. In this way the 'phenomenon' becomes half-appearance and half-phenomenon, and its ambiguity remains an unsolved categorial problem in the history of philosophy.

The third characteristic contradiction inherent in the 'phenomenon'-category is its particular activity. It has the quality of a performance. The 'phenomenon' 'manifests itself'. It not only 'is', but its existential nature challenges; it 'addresses' man and calls his attention to itself. Its phenomenal nature is delineated precisely by this form of manifestation.

[10] Hartmann formulates this 'equalization' as follows: "From the viewpoint of the subject all being-in-itself is transcended (on the premiss of consciousness) into a being-for-me (what confronts me); from the viewpoint of the being-as-being all being-in-itself and being-for-itself is transcended into simply being. The ontological concept of being-in-itself thus manifests itself as the return of the aspect of being from *intentio obliqua* to the *intentio recta.*" (That is, from the epistemological relation to ontology. — M. A.) [N. Hartmann: *Zur Grundlegung der Ontologie, Op. cit.,* pp. 141—142.]

It seems as if a certain element of reality were asking to be admitted to the range of man's attention (his field of praxis) and so were emerging from the background of what has until now been indifferent objects, movements and phenomena. This characteristic feature of the 'phenomenon' is furthermore corroborated by the fact that it is not created by the knowing subject. It is not we who create the hallmarks of reality. On the contrary, we do know that this reality is indifferent to man's endeavours and that it does not reveal its secrets of its own accord. Reality does not help man. It seems as if reality, through its rise to the status of a 'phenomenon', were assuming an active role, but we realize that it would be absurd to suppose such an activity or, at any rate, an active role which would fit in or conflict with the domain of man's teleological practice. Why does the 'phenomenon' have then this contradictory quality of performance, and of what more profound relationship is it the superficial form of manifestation?

The question is all the more important because, as is well known, phenomenology deduces the methodological attainment of its subjectivization of phenomena precisely from this performance quality. In addition, there is a grain of truth in this effort, for indeed the element of subjectivity cannot be wholly excluded from the appearance-category whether appearances are natural or, of a different quality, social. This is also borne out by the endless process of the endeavours of natural science to rid itself of anthropomorphization and it applies even more to the social phenomena where the cognitive-perceptive-active consciousness is not only the 'observer' of the objective nature of phenomena but also an essential constituent of it. Naturally, as we shall see, this role is merely an episodic one. Phenomenology inflates this episodic role in order to prove the entire subjectivization of the phenomenon-category. And the above-mentioned property of the 'phenomenon'-category, performance, weighs as a cogent argument in this evidentiary process since it seems, in fact, as if it were the phenomenon that 'addresses' man. And with Husserl this activity figures in the sense of a piece of reality created from this performance and translated into a 'form of phenomenon' by the subject.

Husserl's starting point is not too far from the actual relationship. The performance-quality of the 'phenomenon' does indeed borrow its form-creating characteristic from the structure of human action. However, it does not do this in terms of the conscious subjectivity but rather as the constituent of a more elaborate process. The reason for this is that in the course of our activity we also bring in motion forces whose effects we do not take into consideration and which, moreover, we do not need during the given action. But these forces, irrespective of our conscious objectives and indeed of our subjectivity, begin to function, and suddenly they 'manifest themselves' as some kind of unexpected effect. It is we who have brought them about and made them 'manifest themselves', without consciously including them in our activity. Consider e.g. the phenomenon of metal fatigue. Properties, which for some time have served the purposes of technological processes and ensured the feasibility of putting the metals to practical uses, suddenly vanish. The metal loses its resilience, its load-bearing capacity, its tensile strength etc. What manifests itself is 'fatigue' — the failure, the particular 'something' that impairs the suitability of the metal. However, this symptom is not yet a 'phenomenon', it confronts us only as a disturbing factor, a contingent case. The repetition of the occurrence provides an increasingly better representation of the series of events of a characteristic aspect, namely that it takes place after a certain period of use as a result of some definite conditions (thermal and mechanical effects).

How does the disturbing factor become a 'phenomenon'? In a way that the disturbing appearances which in the course of the repeated manifestations delineated themselves, a signalling, as it were, their own individualities, for the time being without our knowing their true causes. We are confronted with a signal, with the form of presentation of something which is the manifestation (in conformity with a natural law) of a hitherto unknown coefficient of our own activity. But it is indeed this conformity which we do not know and which we shall even later describe only in ill-defined association with the environmental factors of its manifestation. (Originally, the phenomenon of 'fatigue' had been ascribed to the resultant force of many effects that later proved to be false. For example, the faulty technology of casting, incorrect use etc.)

Notwithstanding that this performance-like quality of the 'phenomenon' has its source in reality, it will not become a signal 'of its own accord'. It is called into being, set in train or, conceived in terms of philosophy, 'premised' by human activity. The 'disturbing circumstance' has been brought about by the concatenation of our actions, by a thus far unknown corollary of their suitably arranged relations of cause and effect. In the course of our struggle with nature, just as in the course of our social actions, we harness its forces, but this means that we also release the kind of energies the effect of which we did not previously have in mind. The reason for this is because teleological consciousness is unable to take in the entire radius of nature's action. As regards our metabolic exchanges with nature, this takes effect in a steadily advancing process. More and more disturbing factors will be divested of the 'phenomenon'-character, to become apprehensible, known phenomena which are therefore 'for us'. In society's metabolic exchange with nature this advance is not always so simple. Certain 'disturbing factors' may assume a character of permanency through social consciousness, finding a 'neutralizing' (practical or ideological) formula by which it can temporarily comprehend even the disturbing factors, i.e. make them a part of the correct action what initially had hindered its teleological practice. To put it differently, our coexistence with the 'disturbing appearances' becomes stabilized with compromise configurations. (This task was fulfilled, for example, by magic.) Gehlen correctly notes that in primitive praxis the way of explaining disturbing circumstances is not to look for their sources but rather to rationalize a certain shock-effect. In his *Der Mensch* [Bonn 1956, p. 331] he writes that the mode of normal elaboration of the disturbances finds expression not in seeking the cause of the disturbance but in a shock. 'To give a wide berth of fire', once you have burned yourself, may be just as effective as, in a different case, the experiment by means of which we pry into the matter of how to treat it so that it does not burn us again. Shocks of this type frequently manifest themselves and are as often as not 'rationalized', i.e., brought to a form of 'certainty' by which they can be explained.

Speaking in terms of philosophy, in the course of action we not only premise our own teleological causal series, but these causalities bring in their train some unintended but objective and necessary consequences. The configurations thus brought into being, the 'phenomena', form the unity (which for the time being cannot be epistemologically differentiated) of the two premisses. This twofold premiss is therefore one of the basic phenomena of praxis, though it chiefly characterizes our metabolic exchanges with nature. In this twofold premiss lies the secret of the performance-like character of the 'phenomenon'-category, the explanation of why it is of the nature both of appearance and of illusion and is indeed a quasi-phenomenon. So let me repeat it again. It is by setting our goals that we also set the 'autonomous' motion of reality in train. But the scope of this motion is always wider than what we can take in and include in our purposeful imagination. Practice is clamped between these two premisses. In certain instances even the opportunity for end-oriented activity is terminated by an unintended consequence. Intensive deforestation has led to the formation of deserts, rendering cultivation impossible in these areas. Here the teleological directedness has led to the discontinuation of itself. Sartre calls this element *'contra-finalité'*, i.e. counter-finality. [Cf. J.-P. Sartre: *Critique of Dialectical Reason,* transl. A. Sheridan-Smith, London 1982, p. 162 ff.] Practice is the unconscious originator of this pair of premisses, since it guides consciously merely one of them. Cognition, on the other hand, joins this process at the meeting of these two different kinds of premisses. The 'disturbing factor' is the signalling of this meeting, bearing out the fact that the suitably planned causal series have been upset. The reason why it 'makes a signal' is because it disturbs our teleological series. Its signal function is given by this deviation from the suitably planned series. It arises from its connection with this deviation. The practical function of cognition consists in recognizing this new group of phenomena and trying to neutralize or put an end to its disturbing effects, or else attempting to make some use of them.

What happens when the disturbing circumstance manifests itself on the side of cognition? In the immediacy of action these 'phenomena' occur at random, so cognition, in order to discover their secrets, wants to produce them only 'as disturbing factors', in order to analyze them and set them in its own causal chains. This is the 'abstract' method of scientific investigation, namely eliciting the 'disturbing factors' in their 'pure' form, independently of the chance events of purposeful activity, and scrutinizing them 'separately'. One model of this method is experiment, repeatable experimentation in which conditions in their 'pure' forms can be exchanged and combined. It is quite another matter that it is hardly possible to produce phenomena in their pure forms. It is peculiar to the ontological complexity of the forms of motion, i.e., to their mode of existence, that they 'overlap' one another and that the ingredients that enter into the composition of any particular 'phenomenon'-structure are constructed also by the determinants of a different range of 'phenomena'.

This is how the explanation for the paradoxical proposition of the 'phenomenon' is found: in the course of his practice man initiates the unknown impulses which later, in the result of his activity, manifest themselves as 'phenomena'. But he observes this result as the self-contained performance of reality, unrelated to the subject. In the immediacy of his view these disturbing impulses get into the scope of action from 'without', independently of him, though he suspects that there must be some sort of a connection between his own performances and that of the mode of manifestation. The 'disturbing circumstance' is still objective, but it is a medium which is alien to the goal, a medium which has nothing to do with our intentions. Should consciousness remain on this level, it would then be in thrall to illusions. But practical necessity urges man to eliminate the disturbance which might render his subsequent life impossible. He is compelled therefore to neutralize the disturbance and discover in the process the common source of his own actions and the 'extraneous' disturbance, i.e., the hitherto invisible effects of the irradiation of subject-being.

Our examples, from technology to the magical treatment of 'phenomena', from the correction of the 'disturbing appearance' to that characteristic of scientific experiments that they rationalize and in fact 'circuit-break' the 'phenomenon'-performance, show even more clearly that the 'phenomenon'-category is, first and foremost, a concept of natural philosophy, notwithstanding that to some extent it is collaterally related to the social forms of presentation and mainly to the active relation of society and individual to these appearances. The collateral relation manifests itself where the individual (or social) performance and the 'independent' premising of reality, over and above the teleological goals, are interdependent in both spheres of action. This twofold premising takes place in a less complicated way in nature where, even if it is through thousands of transposals, the effect originated in praxis goes on reverberating and then elicits an echo in the environmental elements engaged in the various forms of motion, and, creating a new unity of immediacy, manifests itself to the individual, to be ascribed to the independent performance of the 'phenomenon' or to the power of our subjectivity to create phenomena. The same course in human society is run in a more differentiated way by the effects 'irradiated'. The 'mechanical' course of the unaccountable effects, the continuous reverberation and the 'feedback' of response partially obtain here, too. Moreover, in combination with and inseparably from this, the element of consciousness also takes effect. And again through the other individuals' recognition, misunderstanding, acceptance and rejection, the effects irradiated will make their impact and form the answer which stands before us as a social 'phenomenon', as the average of the many different kinds of 'understanding' or misunderstanding reception. Here the 'subjective' side of the performance quality is more differentiated, although even in this case the 'phenomenon' is not to be conceived in the form of a subjective creation or as an arbitrary appearance. 'The road to hell is paved with good intentions.' The truth of this proverb goes to show that an ethical motive (perhaps with positive intentions at that) can easily go to the other extreme when carried into effect, i.e., 'understood' socially, and that it can turn up as an unrecognizable answer, as a 'phenomenon' of a different character, in the form of a 'performance' — a form which we can no longer rightfully identify.

The train of thought of Kantian philosophy and of the school of phenomenology wanted to lay emphasis only on the subjective element in this formula. In other words, it ascribed the performance quality of the 'phenomenon' only to the premising role of cognition: since we are the ones who give things their names and their meanings, it follows that we must have created and moulded them, too. The performance of recognition has become one with the gesture of creation, though in reality the point under discussion is the contradictory unity of the subjective premiss and the objective reaction.

We cannot render the form of origin of the 'phenomenon' independent of the teleological system of action, independent of the manner in which praxis is related to the subject. Whether we are talking about failure, the 'disturbing factor' or the effect of the instruments in distorting our observation, we shall see the suddenly arising 'signal' always in terms of the causal concatenation of teleology, as it is related to the system of our goals. The first form of the stars' motion was given by the practical circumstances of navigation, and the limitations of the state of navigation at that time asserted themselves also in the primitive way in which the stars' motions were viewed. This motion was seen as it manifested itself in the conditions of practice. Though scientific reasoning later broke away from this original form of manifestation, this 'phenomenon'-structure for a long time determined the bounds of both research and the way of seeing things, the outward aspect and the structure of the 'phenomenon'. The long-lasting rule of the Ptolemaic system can be accounted for not only by the prevalence of the religious dogmas, but also by the practical fact that this conception of the universe could give a perfect, virtually accurate course for hugging the coast at the time. (Often a more accurate one even than the Copernican system was to give!) The scope of the practice of the time (hugging the coast, that is) also determined the method of looking at phenomena, but this method could provide sufficient explanation and direction for the practical requirements which gave rise to it. Only later, when sailing on the ocean began, did it appear that the Ptolemaic system was fraught with a very high percentage of errors. Only then did the 'disturbing factors' manifest themselves for the first time. The scope of praxis had widened (i.e., progress had been made in navigation), introducing new aspects of phenomena which had been considered to have been adequately dealt with in both theory and practice. The widened scope of praxis and the true course of the stars' and planets' motions manifested themselves simultaneously in the 'disturbing signals'. And only then could it come to light that the directly visible motion of the stars and planets is merely an illusion behind which there are entirely different laws and motions at work.

The 'phenomenon' is closely enmeshed in the structure of human praxis and cognition, and by raising the level of praxis and through widening its scope and the means available, fresh 'phenomena' will emerge, whereas some others will lose their 'phenomenal' character, changing into 'understood', consequently usable and used, i.e. 'indifferent' phenomena. What the American Indian once saw in the 'phenomenon' of fire was entirely different from what present-day man sees in it, who perceives the factual configuration itself in a different way and in whose brain the spectacle also assumes a different shape. In other words, the primitive level of his practice shows him the phenomenon of fire together with a whole lot of accidental and illusory elements among which the illusory ones prevail, because for the time being it is only through them that he can produce and handle this phenomenon. (That is why fire becomes the most elementary means of magic, and fire magic becomes the first form of conscious social control of this natural process.)

So if we lay such great stress on the phenomenon category's role enmeshed in praxis and on its transient quality in the process of cognition, we must also emphasize the fact that this concept has the characteristics of epistemology rather than of ontology. Phenomenology simply ontologized this epistemological character, just as Hegel's philosophy could not maintain the complexity of the dialectic of subjective and ontological elements when the objective quality of the theory of appearances was discovered, but from the epistemological premiss inferred back to the ontological structure of the sphere of appearances. What with Hegel was still a heroic error, a step towards a more ontological, more objective conception of the universe and appearance, becomes in phenomenology an expression of the uneasiness felt with regard to the permanent revolution of the exact sciences, with regard to modern social illusions, an ideological detour. When we stated above that the 'phenomenon' was mostly a category of epistemology we also meant to imply that in its elements it is the vehicle of ontological aspects but that the limits of its phenomenality and of its 'being so', its recognized, known and used form, are determined by epistemology. Let us take a closer view at this dialectical structure. What we in fact need to do is to throw light on the special structure of 'being for us'.

The simplified idea of 'being for us' eliminates the mediating-immediating role of praxis. In terms of this view it seems as if nature were 'of its own accord' showing us one of its aspects and we were merely appropriating this with the help of our consciousness while remaining in thrall to the delusion of taking the 'for us' form of presentation, that is to say a single part of the objective totality, for the complete whole. However, in examining the 'phenomenon'-category the situation was found to be more complicated. The things' 'being for us' is in fact preformed, prepared and forced into the mouldedness of a certain 'being so' by our 'being for the world', and the unity of these two relations will form the *'So-Sein'* of the 'phenomenon''s 'being for us'. To put it differently, if we proceed from the Marxist category of reality, from the dialectic of the subject-object unity of praxis, we can also obtain a clearer picture of the concept of 'being for us'. The *'für uns Sein'* is also being formed through the teleological premiss. Its so being is also formed by our active, reality-shaping relation which is, however, most of the time unaware of the practical ability of the former to make further impacts. The being for us of the 'phenomenon' developed in this way does not exist outside the relationship of praxis, i.e., apart from our practical performance of shaping. We, too, produce this mode of 'being', 'being for us'. This peculiar mode of being does not exist outside the relationship of praxis. Its true, existential quality is brought about by the interaction in which subject-being (i.e., the active social being), as an individual who can avail himself of material means in a historically determined state of development produces an effect on his social and natural environment and by so doing shapes it, and this reality responds to these effects. 'Being for us' is nothing else but the totality of these objective responses. This interaction shapes the form of presentation of the phenomena, and indeed besides this relationship between subject and object no such 'being for us' exists. The reason why the aspect of the world shown us departs from its 'being in itself' is, as we have already mentioned, because this form of presentation is produced also by our 'being for the world', i.e., by our subject-being realized in our praxis. Hence it is not true that reality is pictured on to a form of presentation 'for us' of its own accord. This presentational configuration is the premising work of praxis itself. While this 'being for us' is an ontological category, it cannot be made independent of praxis, of its premising work which functions teleologically with ontological effects. In terms of this idea we can apply this conception of Marx's first Feuerbach-thesis to the critique of the 'phenomenon'-category. For indeed Marx regards it as a fault of the older materialism that the chief defect of this view is that things *(Gegenstand)*, reality, sensuousness, conceived only in the form of the object, or of contemplation, but not as sensuous human activity, practice, not subjectively. In reality the data of the world are not simply the 'ready' objects facing the subject. Ontologically, in their natural form, they are obviously 'ready' but their quality or shape for us is created precisely by praxis, by the particular practice in which the 'subjective aspect' (the teleological directedness, the ability of the means to impart a shape) and the order of the objective relationships are interwoven with one another. The object is not simply a thing but it is the objectivity which presents itself and assumes a shape in and beyond practice in cognitive process. 'Being for us' is the product of the interaction of subject and object, of their tension and their active and self-forming unity, a product which always exceeds the historical level of cognition. (We shall see a more detailed dialectic of this relation and its true and virtual unity in our analysis of the category of 'object-making' and of its role as a vehicle for illusions.)

Be it said incidentally that 'being for us' simultaneously hypothesizes that the subject also undergoes a process of formation in connection with praxis. In the course of our work we shape not only our objects but also ourselves as well as the notions we form of reality. Consequently, 'being for us' also transforms the self-reflexive inner characteristics of subject-being.[11]

But even if we stress the point that 'being for us' objectively relates to praxis it does not mean at all that we can thus completely ontologize this category. We have underscored merely its relative objectivity, its character of praxis-induced objectification, without which it is not possible to describe even its epistemological data. Should we try to absolutize the ontological quality of this mode of being from its element of objectivity, we would get to the viewpoint of phenomenology, to a subjectivistic portrayal of reality. On the other hand, if we were to deny its objective, ontological elements, we should not be able to understand the progressive dialectic of cognition, united with and yet different in its ways from praxis. We must find therefore the real structural bases of this category between the two extremes. Where is this 'middle' to be found then?

The answer lies in the fact that 'being for us' is to a certain extent still an objective configuration and so, as we have pointed out above, an

[11] According to Hartmann, man, in his work, has an experience of both himself and his object: he experiences himself in the spontaneity of the physical and mental energy he exerts, while he experiences the object in its resistance to this energy. He himself and his object are bound together irrevocably. [Cf. *Zur Grundlegung der Ontologie, Op. cit.*, pp. 205 and 206.]

inferior degree of being, whereas the 'phenomenon', the perceptual apparatus of this relational existence, its conceptual and intuitive form, exists only for cognitive consciousness. The 'phenomenon' exists only in respect of epistemology, whereas 'being for us' which it signals and grasps is still an existential datum in its form which refers beyond itself and which combines many objectively ontological connections and such effects as emanating from subjective praxis. Hence the 'phenomenon' is a more restricted category than 'being for us'; it is the latter category's socially, historically and even ideologically assumed form of perception, its form-creating frame. The basis of the 'phenomenon' and the final driving-force of its 'being for us' is, as we have seen, a quasi-ontological relation, namely the effect which subject-being has upon the object of practice. However, from this relation a form of presentation will arise for a while which is still abstract and so takes form only in an epistemological connection, in cognitive consciousness. Consider our basic example, the 'phenomenon' of the 'rings' occurring round the stars, which primitive telescopes caused to be seen 'on' the phenomena. In this picture the objectively phenomenal presents itself together with the mediating process of cognition, and this mediation is projected on to and identified with the phenomenon itself by the performance of cognition. But as soon as we can examine this same phenomenon in other contexts, too, this intertwined epistemological and ontological aspect will at once separate. It will get rid of the objectification which it has acquired only in the subjective, that is, cognitive process of the 'phenomenon' and we will examine the relationship of 'being for us'. That is why we have said that the 'phenomenon' is a phenomenon which exists only for the subject. Though the 'thing' it signals exists (that is why it can in fact be recognized and made the object of practice and cognition), but only in a form the frames of which are created by the given limits of cognition. And indeed, cognition even identifies these frames (which derive from the reflection) with the phenomenon itself. It objectifies the 'distortions' which arise from the primitive cognitive process, representing as a feature of the objective phenomenon what is actually the distortion of the mediation.

The 'phenomenon' represents the primitive form of the object-subject unity of praxis. In it the 'being for us' of the world does not yet manifest itself as an objective configuration that is independent of consciousness, but merely as a conscious formula. Its objectivity is corroborated by the fact that we have noticed it and interpreted it, that we have objectified it. But in the course of this objectification we have added to it something which, in consequence of the active, premising property of cognition, does not belong to it. Hence instead of grasping data by their 'being in themselves', we can see them only as something else, embedded in and entwined with a series of false transposals and collateral circumstances. And for a while, at the initial stage of praxis and cognition, we must regard as objective the subjective image which is acquired in this way and which persists in the reflection. Only on this basis can we pursue our activities which — likewise for a while — will even be successful. Practice will corroborate these 'phenomenon'-like images. It will corroborate them, because the objective connections are nonetheless present in the subjective elements of the 'phenomenon'. The 'phenomenon' is an epistemologically originated 'husk form' which presents the phenomena wrapped in the historically given praxis possibilities of human development (its means, methods of examination etc.), representing them as objective. In this respect, 'being for us' has a subjective frame: the 'being so' of the 'phenomenon' exists only for our consciousness, though in this 'being so' there is also latent the objective 'being for us' from which we can unearth the true connections once we are at an advanced stage of our praxis. The more ways there are in which we can approach our object (i.e. the larger the surface of praxis in contact with its object), the less significant the 'phenomenal', that is, subjectivistic, frame of presentation will be and the more conspicuous the ontological structure of 'being for us' will become. This is also 'provoked' by practice. It is the answer to the 'questions' practice poses, yet its form of presentation is no longer given by the sole possible transposal of cognition, as in the case of the 'phenomenon', but by the relation to praxis. In the course of the continued expansion of practice this limiting factor will also disappear. We shall not only review and use phenomena with reference to praxis and therefore one-sidedly but we shall be able to cognize them in their relation to other things as well, in the peculiarity of 'being for others'. This third stage gives the phenomenal sphere the frameworks of complete objectivity. If all goes well, this more recent, more clarified stage will not clash with the foregoing one. Only the inner core will come to light, purified of all contingent disturbing, surface elements, a core which has already glimmered in the 'phenomenon' and which may have been even the motivator of correct practice on the basis of the 'phenomenon'. It may also be the case, however, that the 'phenomenon' objectifies an illusory connection which will be made transparent and stripped away as an outward disguise when approached by the progress of science or by social practice (e.g. in the case of wages by the characteristic forms of class struggle) from several different angles. Hence 'phenomenon' is an ambiguous category. It may be the frame of true connections, but it may also be the unconscious expression — the cover-idea — of illusions. And historically this latter case is more likely, for this ambiguity is fraught with peril only in the case when the 'phenomenon' is taken as an objective and real connection, as a phenomenon, and used as such.

This ambiguity stems, among other things, from the fact that, besides a passive receptive (reflexive) role, cognition also has a certain quasi-objectifying capability. As long as we take the configuration of the 'phenomenon' as reality itself it will record these connections as existent which, though they may be real in their components, are nevertheless only virtual in their relations. And rendered defenceless against the sphere of 'phenomena' we must accept this objectively true image as real, because for a while we do not have a different correcting image mediated by our cognitions. It is the epistemological character of the 'phenomenon', its quasi-objectifying function, which determines the two different kinds of possibilities leading to the temporary or permanent setting of the real or illusory configurations.

This is the ambiguity that makes the 'phenomenon' so attractive for modern philosophy. It was born as a crisis category because, by circumventing the unequivocal answer of the ontological structure, one could build a quasi-real, quasi-illusory world with it. The crisis in the development of modern sciences and the 'untrue' factuality of the highly developed industrial society have forced the development of philosophy in the direction of accepting this ambiguity as an aid to devising and ideologically underpinning an unambiguous formulation. These partly scientific and partly ideological requirements covered up the fundamental contradiction of the 'phenomenon'-category and determined its philosophical function.

We have now reached the crucial point in our criticism of the 'phenomenon'-category, the most important conceptual difficulty in the theory of social illusions. What we have said so far has hinted at the contradictoriness inherent in the category, but what has increased the riskiness of using it is that its ambiguity has been generalized and made permanent by a conceptual transposition. The concept of 'phenomenon' had originally been a category of natural philosophy and of sound epistemology, but in modern thinking its validity and its descriptive role were widened. This category is used to describe the social form of presentation and in fact it and its illusions are neutralized. Notwithstanding that there are numerous common elements in the ontology of the presentational forms of both nature and society, the two structures differ from one another in that the social medium is the totality of the conscious, teleologically active people. So the transposition was a rash measure to take, even from the point of view of science. As we have already mentioned, Husserl set about this transposition with great caution. It is true that the theory of values, the analysis of the various meanings and signs of the *Logische Untersuchungen* [1900] and the *Ideen zu einer reinen phänomenologischen Philosophie* [1913], as well as his social references which are sometimes remarkable for the material they offer by way of examples go beyond the strictly scientific scope, but in *Die Krisis der europäischen Wissenschaften und die transzendentale Phänomenologie* [1936] he tried to use more sober concepts. Since, however, his former studies had already initiated this conceptual transposition, the process could not be arrested. Scheler, Heidegger and then Sartre (also the Vienna Circle in a different line of development) use this category explicitly as a social, or even a socio-ontological one, and of course in such a way that the 'phenomenon', together with its subjectivistic and ambiguous characteristic, becomes the objective and ontological synonym for social phenomena. In this concept, as Sartre expresses succinctly, the essence of the 'phenomenon' is presentation, which is no longer opposed to illusion: presentation, at best, is only the measure of illusion. [Cf. J.-P. Sartre: *Being and Nothingness,* transl. H. E. Barnes, New York 1973, p. 4.]

In our following chapters we shall return to the historical connections of this conceptual reinterpretation. Now we will only establish the most important result of this transposition. What occurred was a release of the 'phenomenon'-category in the course of which the illusory and the true presentational forms merged into one another so as to blur their individuality, and moreover precisely this interweaving acquired an ontologizing designation. What at the outset figured as a consciously epistemological (i.e., approximating) resource became artificially ontologized in this subjectivistic form. This interweaving was not grasped consciously. It was with a naive ease that the 'phenomenon'-category has become widely known as the in some way common category of reality and of thought. Its apparency is hidden by its ontological role, and even today it is treated as the synonym for Hegel's 'appearance' and indeed for the Marxist idea of 'presentation'. Its inherent contradictions and ideological implications have disappeared, thus making it simply impossible to explore philosophically the social flora of illusions. The initial step was brilliant: to apply the conceptual structure of an inferior form of motion to the apparatus of a superior structure of appearances, to make a problematical mode of presentation into the mode of expression of a real sphere of phenomena. This was the decisive, however negative revolution of philosophy. Hence the narrower category *a priori* eliminated the more perplexing contradictions inherent in the structure of the socio-ontological presentation, it eliminated the set of appearances related to individuality, to the collision between individual and social consciousness, to the praxis on the level of society as a whole, to the developmental paradox of man's species-being, and by means of this simplification it also frustrated all those appearances which would have resulted in the description of illusions, in the conceptual revelation of virtual reality coming about under the shadow of and combined with the true presentation. Purely scientifically, the 'phenomenon' precluded this ontological question-mark, for the overall power of the concept, its penetration in depth simply did not reach this level.

That is why we have had to devote such a lengthy criticism to the structural shortcomings of this category and its inherent contradictions. We had to clear the way for unravelling the problem of illusion. We had to elucidate a philosophical misunderstanding so that we could make room for an analysis of the objectivity and virtuality of illusion. As a matter of fact we had to go back to the range of problems which we examined as 'being for others' in our previous chapter, problems in which we could see the objectivity of presentation, the criterion of its existence independent of the subject. This criterion was lost in the case of the 'phenomenon' when it became an ontological concept in the form of its presentation only 'for us'.

1.3. THE SEMANTIC CHANGES OF A CONCEPT

It is a well-known characteristic of the history of concepts that they sometimes begin a new life with an opposite content. The classical doctrine of forms later became a means of describing contents. The 'realism' of the Middle Ages was the conceptual framework of idealism, hence of a content contrary to the category of realism in use today. Such phenomena of conceptual reincarnation can be explained by the intellectual moment of inertia. Progress, instead of elaborating a new concept, tends to use the traditional categories even when it is compelled to change their content into a contrary one. The evolution of the 'phenomenon'-category has suffered a similar change in meaning. We have already made it clear that we deliberately use this category in a sense other than in which it occurs in the history of philosophy and particularly in modern philosophical trends. In the sense in which we understand it, the 'phenomenon'-category is the appearance-category of the scientific cognitive process, and in the course of the history of thought it has figured as a vestigial concept of natural philosophy and then as its unintentional and undetected residuum. But it is also true that besides the epistemological characteristic a certain ontological semblance was also added to this natural philosophical origin and, moreover, that in the modern view it was mixed with the attempts to grasp categorially the social forms of presentation and social illusions. That is to say, despite the natural philosophical and quasi-ontological origin of the 'phenomenon', it was transferred into a socio-ontological medium, and it is to this inadequate role that its immanent contradictions are due. By way of illustration, we propose to examine more closely three points at which this contradiction-fraught range of roles and the development of this semantic change are interconnected: the Greek solution, the answer of classical German philosophy and the more recent concept-change of phenomenology. While with Aristotle the development of this concept was brought about by the social crisis of the democracy of the *polis* and by the concomitant change in the concept of nature, in Kant's and Goethe's argument the inconsistent character of the concept — pregnant at the same time with different kinds of dialectics — was produced by the sudden crisis of the emergence of modern sciences. In modern phenomenology it was developed into an ambivalent category again by a social-ideological necessity and a more recent developmental crisis in scientific thinking, making it an universal concept which will become simultaneously the medium of illusions. (It is no mere chance that Husserl's most profound study, *Die Krisis der europäischen Wissenschaften und die transzendentale Phänomenologie*, a late work, is written on the methodological crisis of the modern sciences, but even his first important work, the *Logische Untersuchungen,* is a methodological essay but on the lines of natural philosophy.)

We propose to contrast these three interconnections with one another. The category has pretty much the same content in all three versions of the 'phenomenon', but its meaning has become self-contradictory in the course of development. In all three cases this category indicates the inseparability of appearances (phenomena) and illusions, the indissociability of ontological and epistemological relations. It is the product of this interwovenness. But in Greek philosophy, as we shall soon see, this undecidedness is characteristic only of the initial stage of reflection. The reason why the objectively phenomenalistic and the subjective distortion are mixed for Aristotle is because he is still some distance away from discovering the dialectic of appearance and essence, this category being only a single milestone on this road. The cognizability of the world is no problem for Aristotle who sees blank spots only in the structure of reflection. The 'phenomenon'-concept of the 20th century uses this mixed unity to deny reality and to prove the theoretical impossibility of cognition. Since appearance and illusion are theoretically inseparable from one another, cognition always travels about in some precarious space. Husserl and Heidegger work in the downward branch of the history of philosophy, beyond the classical solutions of the appearance-theories. They see their task in demolishing the edifice of rational explanation. Appearances and illusions are inseparable from one another both theoretically and ontologically. They have put on (irrationally) each other's disguise. As a result, we have to face the paradoxical situation that formally Heidegger refers with good reason to Aristotle when he supports the introduction of the 'phenomenon'-concept with propositions from the *Metaphysics.* But this formal identity conceals differences not only of content but also of directedness. These differences are more difficult to discern than, shall we say, the contentual differences of the aforementioned changes of concepts. While the whole content of the concept was changed in the former, only its function, its position in the conception of the universe was altered in this instance.

Heidegger makes quite clear why he proceeds from the 'phenomenon'-concept and why the more up-to-date 'appearance'-concept is of no use for him. His reason is because appearance is an unduly material and objective configuration: "'Phenomenon', the showing-itself-in-itself, signifies a distinctive way in which something can be encountered. 'Appearance', on the other hand, means a reference-relationship which is in an entity itself." [Heidegger: *Being and Time,* London 1962, p. 54.] In other words, what makes this concept suitable for his philosophy is that here the world shows itself only in the conceptual relation of the presentation 'for us', and the image of reality that the subject obtains depends merely on his particular position. By this means we can avoid the problem of the objective existence of appearance, the question concerning the objectivity of the world. "If, however, the phenomenological conception of phenomenon is to be understood at all, regardless of how much closer we may come to determining the nature of that which shows itself, this presupposes inevitably that we must have an insight into the meaning of the formal conception of phenomenon and its legitimate employment in an important signification." [*Ibid.*, p. 55.] It is therefore quite unnecessary to deal with the more detailed contentual determinatenesses and so the question whether the 'phenomenon' is an objective configuration or one of reflection simply does not arise. This must be parenthesized in the same way as the objectivity of exteriorization with Husserl is also but an incidental matter. The 'phenomenon' in this sense represents a picture which has been taken cognisance of, which has manifested itself for us and the development of which has been continued in reflection. Since the appearance-category would not tolerate this subjectification, we must fall back on a still inchoate concept in which these two elements (objectivity and reflection) can coexist still in their naive and original unity.

Sartre formulates these requirements and the functional role of the 'phenomenon'-concept even more clearly. For him the phenomenon itself is also a subjective configuration and so it is identical with illusion. "The obvious conclusion is that the dualism of being and appearance is no longer entitled to any legal status within philosophy. ... But if we once get away from what Nietzsche called 'the illusion of worlds-behind-the-scene', and if we no longer believe in the being-behind-the-appearance, then the appearance becomes full positivity; its essence is an 'appearing' which is no longer opposed to being but on the contrary is the measure of it. For the being of an existent is exactly what it appears." [J.-P. Sartre: *Being and Nothingness, Op. cit.*, p. 4.] The 'phenomenon'-category is the expression of this immediate identity or, as Sartre says aptly, is the measure of it. The idea of identifying appearance (phenomena) with illusion could not have been expressed more clearly. We can see at the same time the collapse of that enormous intellectual edifice erected by the development of thought over thousands of years, a development which brought about both the separation and the unity of the spheres of appearance and essence, the dialectical relationship to which now Sartre — precisely with Nietzsche's words — refers scornfully as day-dreams. Whereas the progressive stages of development of thought have always been at war with illusions dazing the mind, here acquiescence and compromise are given emphasis in the form of illusion: life itself as we live it and as it is at all possible to live it. There is no other world, only the range of illusions exists. With this the struggle against illusions, a struggle which incidentally Sartre also takes on in his literary works, has lost its philosophical cogency. It is this requirement by which we can understand why he had to reject Hegel's concept of appearance and why he had to go back to a formally similar category of Greek philosophy instead.

This critical remark would naturally be biassed if we did not refer to the fact that the method of phenomenology has nevertheless attained important results precisely by the aid of this reduction. We are in the habit of speaking rather disdainfully about descriptive works, but, since the exact description of certain appearances has itself become open to doubt in the current stage of science, the method of phenomenology has turned out to be successful. In terms of philosophy, it has helped the individualization of appearances, the endeavours which were inherent in conceptual apprehensions and opposed to hard-and-fast generalizations. It has furthermore contributed to the continued development of the modern conception according to which substantive relationships are forms of motion of aggregate appearances. The description, as a procedure which in terms of methodology is capable of being rendered independent, could have here, too, a positive effect, regardless of the ideological background of the parentheses. This philosophical attitude also played a part in, for instance, the propagation of statistical methods. Even Marxism can learn something in this respect, as regards the dialectic of the individualization of appearances and of their universality which is to be apprehended in that individuality. Husserl's phenomenology took important steps in the direction of this dialectic, even if from an ideologically incorrect starting-point. One of the major difficulties in the Marxist school of thought today is the generalizing and universalizing conception of appearances, a conception which, by skipping the singularity of socio-historical configurations, works with universal categories and so is unable to grasp the particularity of appearances.

Identifying appearances with illusions is not the discovery of phenomenology. Sartre's reference to Nietzsche indicates that here we have to deal with a more sweeping trend. Phenomenology gave only a name and a categorial explication to the endeavour which involves the entire bourgeois philosophy of today. Thus e.g. Moritz Schlick, the founder of the Vienna Circle, one of those who imparted a new impulse to neo-positivism, writes that it is easy to realize that to draw a distinction between a thing and an appearance is rather impractical and that sooner or later the appearance-concept must disappear from philosophy. In connection with the analysis of the Kantian noumenon he sums up the new situation in the following terms: We reject the Kantian concept of appearance with respect to these ideas. Our experiences and our perceptions, our imaginations and our feelings are not epiphenomena, they are not merely appearances, but they are independent realities to the same effect as all other transcendental 'things' are. There is but a single reality and this reality is always essence and cannot be separated into appearance and essence. [M. Schlick: *Allgemeine Erkenntnislehre,* Berlin, p. 216.] Neopositivism denies the Kantian, essentially antidialectical concept, and not Hegel's more developed concept of appearances, and it does so by criticizing its unstable and thus properly contestable theory from the right. Everything which presents itself is essence. In simplified terms, this is the final conclusion of neopositivism just as it is that of phenomenology. But here the category which could sum up this universal essence does not yet exist. The 'feeling', the 'perception' or 'experience' can be used that much less the more they betray their subjective origin. The quasi-objectivity of the 'phenomenon', the contradiction of its manifestation which we have already analyzed, namely that, although its 'husk' is given by the subjective scope of cognition, it nevertheless has objective elements as well, makes it a much more advantageous category. This type of subjectivistic conception of appearances arises in a less refined or more subjective form with Hans Vaihinger [*The Philosophy of 'as if'*, transl. C. K. Ogden, London 1965] also in pragmatism, semantics, and more recently in the linguistics of structuralism. But the theoretical bases of this conceptual attitude were laid down by phenomenology, so it is no mere chance that, directly or indirectly nearly every modern trend of philosophy has since proceeded from this doctrine.

Now we may take up the question of the extent to which we are justified in ascribing this concept to our Greek forefathers and primarily to Aristotle's concept of *φαινομενα*. The relationship between the immediately given reality and the hidden 'permanence', the problem inherent in the reality of the phenomenal world had appealed to Greek philosophy practically from the beginning. This problem returned time and time again: with the Eleatics as the reality of kinetic phenomena, in atomism as the discovery of the essence of reality, in the Platonic doctrine of Ideas as the denial of the objectivity of immediate appearances. From among these wide-ranging problems we shall examine only the structure of Aristotle's concept of *φαινομενα*, this being the most important attempt at a solution. Aristotle continues the chain of thought which has been in search of the dialectic between illusoriness and the contradictory nature of motion, from Heraclitus through the Eleatics and the Pythagoreans all the way to Plato. The contradictoriness of motion manifested itself not only as a problem of dialectics but also as an ontological question-mark. If motion itself is contradictory then either (1) the appearance is itself false since the essence and the truth of the world are free from contradiction, or (2) the appearance is true, but then the whole of the reality is also of a contradictory structure and ontology must be constructed accordingly. The Eleatics, the Pythagoreans and to some extent Plato himself set off on the first road, while Heraclitus and Aristotle attempted to find a solution on the basis of the other version. However, both methods were conducive to additional contradictions. Only the Eleatics conceived the theory of the apparency of kinetic phenomena with a fine-drawn distinction by naively assuming the coexistence of the motionless world and the illusory motions. Plato, on the other hand, discovers that contradiction is a more comprehensive category, and though his doctrine of Ideas is not imbued with this tension, this does not interfere with his explaining the whole of reality by the dialectic of becoming. With this, however, there arises a contradiction between the restful world of ideas and the visible reality seething in the becoming, a contradiction which Plato is unable to resolve. This visible world is a 'reflection' of the ideas, and yet there occur contradictory appearances in the carnival of the mirror-images, whereas the world of the original 'reality' (of the ideas) is free from contradictions. Hence the system is unstable.

The Heraclitean legacy, the attempts of the materialistic dialectics, also got involved in contradictions. The world in motion is reality and not false appearance — but if such is the case then what is it that moves this structure, or rather what is behind the kinetic phenomena as a substance, as an ultimate driving force? If the driving force is restful then the system will be upset again, for indeed no motion can arise from rest. On the other hand, if the substance itself is in motion then what distinguishes it from the world of phenomena, from the sphere of the configurations of motion?

Aristotle's greatness as a thinker is shown by the fact that he cut a way through this jungle of problems and sought the solution in the right direction. As contrasted with the intellectual convention from which Greek philosophy could barely depart, namely that appearance is a lie, that it is an error, Aristotle surmises that reality itself is hidden in this lie. He formulates the concept of *φαινομενα* several times and in contradictory propositions, but the point under discussion is always the same in these conflicting formulations: the erroneous element of reflection and the 'contingent nature' of the real phenomenon, its not-quite-true characteristic are combined with each other in the 'phenomenon', and so the element of reality and of truth continues to subsist in it just the same. While the noted and frequently quoted analyses concerning the perception of colour in his book *On the Mind* lay stress on the subjective aspect, in his *Metaphysics* he comes nearer to the objective appearance-concept. The 'phenomenon'-category is 'on its way' to point out the outward relationships of reality.

The conflicting formulations in the *Metaphysics* are, however, due not only to Aristotle himself also being undecided about summarizing the content of this concept, emphasizing now the subjective now the objective, but also to the fact that in the various historical strata of the genesis of the *Metaphysics* this category emerges in different forms. In the earlier, 'mature' chapters [Books A, B and C] he grasps the 'phenomenon' from the subject's point of view, as a result of the relative nature of cognition. "Regarding the truth, we must maintain that not everything which appears is true, firstly because even if sensation — at least of the object peculiar to the sense in question — is not false, still appearance is not the same as sensation." [Aristotle: *Metaphysics,* transl. W. D. Ross, Oxford 1923, 1928^4, 1010a.] Here the 'phenomenon' is identical with the thing given in sensation. But already in this context he establishes that the phenomena which are obtained in this manner are at the same time irradiations of the substance and that they can be separated from the illusions precisely in this reference to their contents: "For sensation is surely not the sensation of itself, but there is something beyond the sensation, which is necessarily prior to the sensation; for that which moves is prior in nature to that which is moved." [*Ibid.*, 1010b.] Then he adds: "Yet all these views [the Platonic doctrine of Ideas — M. A.] destroy this necessity, leaving nothing to be of necessity, as they leave no essence of anything." *[Ibid.]* Hence Aristotle clearly defines the conceptual contents of the 'phenomenon': it is, on the one hand, the uncertain presentation of the substance, and on the other, the distortion of cognition. Both factors can produce the 'phenomenon': either the substance as an outward form, or the sensory delusions of cognition. Aristotle illustrates the point by the change in the sweetness of wine, a change which can be due to either of two different reasons. Either the wine itself undergoes a change or man's capacity of flavour reacts differently (e.g. one would find the wine sweeter if one had earlier eaten something sour). But sweetness, as a phenomenon, is still a quality which originates in the things themselves. ('But at least the sweet, such as it is when it exists, has never yet changed...') Aristotle conceived a contradictory category, namely that the 'phenomenon' contains both the subjective element and the objective phenomenal nature of reality. Hereby he clearly draws the line between himself and the Platonic view. For indeed, the root idea of Plato's *Theaetetus* is that appearance is merely the category of perception and that, in this respect, it is deceptive, contingent only on the subjectivity of the sensory organs. [Cf. Plato: 'Phaedo', in: *Collected Works,* Vol. 1, transl. H. N. Fowler, London—Cambridge (Mass.) 1953.]

The two contents still coexist in these earlier chapters of the *Metaphysics.* Besides emphasizing the objectivity in these books he lays equal stress on the subjective character of the 'phenomenon': "... but that which appears exists for him to whom it appears, and when and to the sense to which and under the conditions under which it appears". [*Ibid.*, 1011a.] The point he emphasizes now is the relation of sensory cognition.

As opposed to this wavering attitude which only predicates but which fails to reconcile the contradiction of the content, Aristotle defines the structure of the 'phenomenon' more objectively in chapter 'S' of Book L, an insertion he made only at some later time. However, this 'more objective' formulation also bears the inspiration of the progress of ancient natural philosophy, although it also results from the development of a social crisis. (This natural philosophical origin is demonstrated by Jaeger in his analysis to be outlined below.) It was Jaeger's researches, authoritative even today, which called attention to this difference in time in the construction of the *Metaphysics.* [Werner Jaeger: *Aristotle,* transl. A. Robinson, Oxford 1934.] Jaeger discovered that in this later solution Aristotle was already inclined to assume separate prime movers, independent substances behind each phenomenon. He had earlier taken the view that there was but a single mover sustaining the world, and just as the Eleatics or Plato's doctrine of Ideas he had also been involved in inextricable contradictions.[12] The theory of the many 'individualized' movers can provide, however, a more rational explanation not only for the phenomena of motion. If there is a separate mover behind each phenomenon (stellar motion), then the 'phenomenon' will function in the sense of an objective phenomenon. The hitherto insoluble problem posed by subjectivistic sensory delusions is resolved. Of course, with this new hypothesis thrown up by contemporary astronomy and primarily by

[12] In connection with this conceptual development Jaeger writes that it was evidently the specialized branch of science, i.e. astronomy, which gave the impetus to improving the theory of the prime mover. This scientific accomplishment directed Aristotle's attention to the fact that the hypothesis of the sole primordial mover was too simple and that it could not account for the real motions of the stars. Jaeger adds moreover that Plotinus already noticed that he was confronted here with a new formulation of the concept of appearance. The only thing he objected to was that the 'independent' substances behind the appearances would upset the harmony which the sole primordial mover had still been able to create. But this dialectical problem is insoluble for both Aristotle and the Greek philosophy.

Eudoxos's theory, several new contradictions arose which Aristotle was unable to resolve. There is nevertheless one thing that can be seen clearly, namely that he sought the dialectical solution of the problem of appearances in the direction of objectivity. He defined the 'phenomenon' category as something aiming for reality. The only way in which we can understand this formulation including a dual content is if we examine his experiment in the light of this new trend of development. Aristotle here swept forward towards Hegel's concept of appearance, towards the discovery of the objectivity of the phenomenal world.

When phenomenology desires to utilize this ambiguity in order to fulfill its own subjectivistic requirements, then formally, abstracting from the historical contexts, it proceeds correctly, for the *Metaphysics,* considered as a coherent work, is in fact fraught with contradictions, and indeed it asserts a position replete with ambiguous formulations with regard to the concept of 'phenomenon'. However, this ambiguity, in the matter of Aristotle's development and the trend of his researches, points towards the direction of unambiguity, towards the clarification of objective phenomenal existence. But phenomenology does not want to take cognizance of this historical relationship, for essentially it identifies Aristotle's point of view with the attitude of the late Greek philosophy, with the theory of Pyrrho and Plotinus, where it is indeed the subjective element, that is, apparency, which prevails as opposed to the objective elements of phenomena. According to Pyrrho, each person has a different visual perception of the 'phenomena', and there is absolutely no common element in this dissimilarity. Pyrrho, according to Diogenes Laertius, writes that "we admit the apparent fact without admitting that it really is what it appears to be". [Diogenes Laertius: *Lives of Eminent Philosophers,* transl. R. D. Hicks, Cambridge (Mass.)—London 1958-59, Vol. II, p. 515.] The phenomena of the disintegration of Greek philosophy such as e.g. the freezing of the elementary initiatives of scientific thinking and thereby the speculative reconstruction of natural philosophy brought in their wake a series of new dialectical problems, including the question of the objectivity of illusion. Phenomenology, in search of the primordial things, is nevertheless not inclined to emphasize these problems and regards this development only as a sequence of endeavours from the subjectivity of which Aristotle's viewpoint can be understood. Thus the original content of the concept of 'phenomenon', namely the 'directedness' arising from the development of Greek philosophy is lost, and thus there remains only the bare and unmistakable 'ambiguity': the identity of 'phenomenal' illusion with the phenomenon.

The possibility of establishing a modern interpretation of the 'phenomenon'-concept was afforded by the aid given by the revival of the Kantian conception. Besides the up-to-date requirements of ideology the antinomy of phenomenon and noumenon was the conceptual base on which the category of the unity of appearance and illusion (the distorted theory of the 'phenomenon') could be elaborated. Kant, by separating appearances from essence and setting them in two separate spheres, simultaneously afforded the possibility that no distinction could be made between appearance and illusion. If there is no yardstick by which we can compare the two then theoretically we cannot see the essential connections, and the phenomenal world in its own superficial relationships can equally be illusory and expressive of reality. Consequently, though Kant did not draw these inferences yet, the ambiguity of the 'phenomenon' and the unity of its subjective and objective elements could be maintained and, moreover, the leading role of the subjective elements worked out.

But in order to understand Kant's viewpoint we must bear in mind that with the intense development of the exact sciences there had emerged numerous dialectical problems which nearly without exception add up to the strikingly increasing separation of appearance and laws (regularities) that tend to become increasingly distinct spheres. The various branches of science have directed attention to an increasingly greater number of relationships where the directly visible structure and the real one are in sharp contrast with one another. Consider e.g. chemistry where there is, as it were, an unbridgeable gap between the experiential chemical phenomena and the actual chemical reactions which can be expressed in formulae. It was too great a 'shock'. The shock of the discovery which has created an increasingly greater distance between the directly visible and that which is lawlike, especially in the different branches of modern natural science such as chemistry, biology, electricity, thermodynamics etc., that is to say in sciences which function with a spectacular phenomenal sphere rather than with intuitive categories and laws. The empirical husk of the phenomenal world and the scientifically expressible law could seem to constitute two different planes. This also gave rise to the need for this shock to be formulated as a problem. Earlier, dialectical contradictions had manifested themselves only in their germ. Now, however, the progress of science suggested the unreliability of the 'phenomenon'-structure and its constantly changing and therefore 'false' nature. It also came to light — precisely as a result of the scientific revolution — that the relationships which had earlier been supposed to be lawlike were to be replaced by more recent laws, and this indicated that the 'essence', the 'lawlike', itself tends to hide itself, and one can perhaps only fumble in the dark, trying to help oneself by having recourse to ever newer laws, without being able to reach the real thing as it is in itself. Kant gives this scientific crisis-consciousness a philosophical form. Driven by this shock effect he seeks the answer to these dialectical relationships, and that is why he is unable to elaborate the intricate dialectic between separation and unity. The only conclusions he can draw from this crisis are the subjectivity of appearance and the irrecognizability of essence. The unity of relationship and dissimilarity will subsequently be solved by Hegel through his theory of the objectivity of appearances, and will formally be cleared up by Marxism by separating the objective appearances from the illusory objectifications of a subjective origin. Here there can be light thrown for the first time on the differences between 'phenomenon' and appearance. Kant is still at the start of this development, and in his view the 'phenomenon'-category represents the entwined unity of two endeavours. He would like, on the one hand, to grasp the dialectic of appearance and essence as it is offered by modern natural science. On the other hand, he attempts to find the answer to the question of how it can be that the development of scientific reflection advances ever deeper, and that the relationships which had earlier been thought as laws will again give rise to appearances or even to errors. These two problems are entwined with him in the concept of 'phenomenon', and since he cannot separate the epistemological contradiction from what is already an ontological structure, there will arise the antinomy between the 'thing as it is in itself' and the phenomenal.

It is characteristic of the bourgeois thinking of today that rather than continuing the achievements of the heights of development and enriching Hegel's dialectical ideas, it turns back to Kant and approaches the entire history of philosophy on the basis of his rudimentary and inescapably agnostic standpoint. Aristotle is also explained on this basis, and the modern theory of 'phenomena' is also worked out within the framework of this philosophical outlook. Hence Lukács is only too right in writing that to evade the dialectical discussion of problems is conducive to irrationalism.[13] What with Kant was an unsuccessful and interrupted attempt at elaborating the dialectic of appearance becomes later, as an escape from dialectics, an auxiliary idea of irrationalism; reason is utterly helpless in the face of appearances and is unable to decide what is illusory and what is essential.

[13] G. Lukács: *The Destruction of Reason,* transl. P. Palmer, London 1980, pp. 103—105.

We shall select merely a single point in the post-Kantian development, as the rearguard action which points in the direction of both the defence of the objective theory of appearance and the evasion of a crisis caused by the natural sciences, without reaching Hegel's solution. This point is Goethe's theory of *'Urphänomen'* or primordial 'phenomenon'. Goethe is also a witness to and, to some extent, even a creative participant in the scientific crisis which by denying the anthropomorphic wealth of the immediate phenomenal world gets to the increasingly more abstract and fast-paced 'changing process' of laws and regularities. (Consider e.g. his colour-theory and his anatomical investigations.) Hence he is intrigued by the same problems with which Kant was also concerned, namely with the 'uncertain' nature of the phenomenal world. Yet, he disagrees with the attitude of the *Critique of Pure Reason,* for neither as an artist nor as the last representative of natural philosophy can he accept the denial of the phenomenal world. That is why Goethe elaborates his theory of *'Urphänomen',* his rearguard action against the disanthropomorphizing endeavours of modern scientific development. He develops for himself an already anachronistic position of natural philosophy, though, true enough, he does so with little enthusiasm and rather in defence of the artist's contact with reality. The point under discussion is essentially to defend the phenomenal and the structure which renders visible the laws of the humanly perceptible world. Here, too, there can be seen the fundamentally natural philosophical origin of the 'phenomenon'-category and its character which, running contrary to a theory of appearance which describes social relations, results in illusions.

Inspired by this fundamental purpose Goethe comes upon many dialectical problems which in their own way, in a roundabout way, will later help to clear up the appearance-concept. His is not a personal argument with Kant. Goethe does not even mention Kant's name, but from the context it is clear that it is to Kant that he is linked by the remarkably exciting philosophical problems of the period. Goethe formulates his theory of *'Urphänomen'* twice, first in 1793 *[Erfahrung und Wissenschaft],* a few years after the publication of the *Critique of Pure Reason,* and then in 1825 *[Versuch einer Witterungslehre],* this time after the investigations of Hegel's *Logic.* The first study proposes to surpass the Kantian conception only in terms of methodology. The reason, in Goethe's opinion, why appearance and the 'ultimate' essence are not opposed to each other is because 'pure appearances' need only to be raised out of the sea of relationships of the 'phenomena'. Thus, for instance, a tree has many different kinds of leaves, but we can still establish the characteristic and lawlike pattern of the leaves of this particular tree. And likewise, though the individual members of animal species also vary, yet the 'pure type' can nevertheless be constructed to a fault. The only thing that Goethe fails to notice is that by describing pure appearance we have still not provided an answer to the question of what sort of relationships might bring about these appearances, keep moving them etc. On the other hand, he gives an excellent description of the phenomenal, distorted in the process of cognition and subjectively discoloured. "For, since the observer never sees the pure 'phenomenon' with his eyes but much depends on his state of mind, the state of the organ (eyes) at the moment, light, air, weather, bodies, treatment and a thousand other circumstances, it follows that a sea must be drunk up if one wishes to retain, observe, measure and weigh the individuality of the phenomenon. ... If I have experienced (learned) to a certain degree the constancy and consistency of the phenomena I deduce therefrom an empirical law and prescribe it for future appearances. ... This then according to my experience would be the point at which the human mind can most approach objects in their universality [and] draw them to itself..." [Verbatim translation. Goethe: *Naturwissenschaftliche Schriften,* Weimar 1893, II Abt., Bd. XI, p. 38. — Hereinafter referred to as *Naturwiss.*] The 'pure appearances' obtained in this manner are laws, too, and in this 'archetypal' form these two cannot in a Kantian way be separated from one another any more: appearance and essence are inseparably united in the *'Urphänomen'*. In one of his later remarks he still seems to be clearly engaged in debate with the Kantian view. He proceeds from the idea that although appearance has become separated from essence but will nevertheless take a new, independent and essential form. In his study, 'Der Versuch als Vermittler zwischen Subject und Object' [in: *Naturwiss.*, p. 21] he writes the following: What presents itself must separate so that it will be only an appearance. But that which has been separated will again be in search of, and form a unity with, itself. In its inferior forms this encounter may be a simple mingling when appearance is null or an indifferent thing. If, on the other hand, that which has been separated intensifies itself and enters into relations with the other in this intensified state, some sort of a new thing, a third something will arise. Here Goethe is on the point of surmising the unity and difference of subject and object, and in describing the structure of this experiment he intends to evade the constituents of this unity as the crisis symptoms inherent in the appearance-concept of modern natural science. Here, too, he fights against the Kantian antinomy: though the existence of appearance breaks away from essence, yet it will form a new unit with it. Hereby he already displays a presentiment of Hegel's concept of appearance. Goethe focusses his investigations not on the 'phenomenon'-category, but on appearance. And he also knows that this appearance must first be produced separately, in its 'pure' form, rid of its anthropomorphic contingencies; this is an objective which is served by the artificial and abstract medium of the experiment. But pure appearance, i.e. the 'intensified' 'phenomenon', happens to be the essential relationship itself. As against the post-Hegelian conceptions tending to simplify matters, here there arises a genuine problem, namely that essence 'manifests itself' not beyond appearances but in appearances themselves. But the trouble is that the mode of existence of this sphere is not yet clear here. Goethe believes in the illusion that the noumenon-sphere can be eliminated.

Goethe tries to break through in the direction of the dialectic of appearance and essence but also to maintain the many-coloured nature of the anthropomorphic world as the underlying material of his artistic character. This new concept, as Lukács points out, will paraphrase the category of particularity but by the nature of things it will also seek here to reconcile the contradictions inherent in the appearance-concept.[14] According to Hegel's solution, as we shall later see in some detail, appearance is identical with essence, yet this identity can arise only by virtue of the difference and can be restored only as a process. While Kant stresses only the element of difference, Goethe restores unity by not letting the abstract, general characteristic of essence take effect. He feels that if he postulates the universality of essence it will be impossible to assume that the two are identical. The universal character is capable of subsuming the individual and particular appearances only in a speculative way at best. In the sense of ontology there are only appearances and within this sphere there arise those intersections (pure appearances) which can be, in their particular forms, appearances and essences simultaneously. The universality of the essence may exist at best in the sense of 'intensification'. We can distinctly recognize in this conception the artist's attitude, the method of typifying appearances. But the hitch is that Goethe proposes to transfer this method on to the field of natural sciences.

In his later study the theory of *'Urphänomen'* arises not only as a conceptual method but its ontological character is also emphasized. The

[14] Gy. Lukács: *A különösség* [Particularity], Budapest 1957, p. 119.

keynote is formed by his famous idea to which we have previously referred, according to which 'nature has neither kernel nor shell, for it is simultaneously both'. In the *Naturwiss.* he writes that nothing which is or which presents itself, lasts or disappears can be conceived isolated in its own bareness, for a thing is always pervaded, followed, clothed and comprehended by another. Things are affected, they are brought about by causes, and if so many essences affect one another then where can we look for that transcendental intuition which might be able to decide which is dominant and which is subordinate? So there is nothing else left for us than to stop somewhere in the middle and to handle the rest as peripheral from this point. Here Goethe is inclined to regard the separation of appearance from essence merely as a process of cognition. It is reflection which breaks up the unified whole of reality into something which is obvious and something which is deep-seated. But at the same time, by virtue of the concept of inherency, namely that things exist in one another and that they can function only in their complex and mutual relationship, he introduces a category of the existence of appearances into the general knowledge of philosophy, a category which we shall be able to really appreciate only as the ontological, or more exactly, materialistic correction of Hegel's solution. This is precisely the point where Hegel's theory is inappropriate. For indeed, with Hegel, appearance as an ontological performance is just as natural and matter-of-course a process as the hiding of essence, its objectification and its loss in the material world. Goethe gives the ontological key to this hiding but he does so in a formulation which simultaneously refuses to show him that there is a unity between the identity and difference of, on the one hand, the transcendency of appearances and, on the other hand, the essential immanence. He is both post-Hegelian and pre-Hegelian.

At some later stage, however, he puts this idea in a positive form: The transcendental attitude would be, he writes, if we admitted that anything factual is also a theory in itself. The blue sky reveals to us the law of cosmos. But we must not search behind appearances, for indeed appearance is a law unto itself. Besides expressing nostalgia for the anthropomorphic outlook, here he also asserts his more general philosophical debate with classical idealism. He is simply unable to accept the premiss that there is a difference between appearance and essence, and he expects to solve the problem in the immediate identity, although he has already expressed his opinion that such immediate identity is plainly impossible, for indeed one faces the accidental 'sea' of appearances, a sea from which one must raise the 'pure' 'phenomenon'. So the concept of *'Urphänomen'* points, on the one hand, in the direction where Hegel's solution will be conceived, where appearance separates from and becomes identical with essence only in the course of mediations. But on the other hand, he also draws the line between Hegel and himself, inasmuch as he stresses the sensory wholeness of the phenomenal, the phenomenal mode of existence of the essence, and cannot imagine that the laws may differ 'in their structure' from 'pure appearances'. And he is likewise unable to realize that with his theory of 'phenomenon' he cannot mark off his conception from the world of illusions since in Hegel's concept the immediate coincidence of appearance and essence is identical with illusions. But for Goethe, who happens to be a basically optimistic ideologist of a revolutionary period, illusion is still of minor importance, so much so that it might as well be left out of his consideration. It is to be noted that this is also the case with Hegel. His concept of appearing is not identical with the doctrine of false illusions, because he inquires only into the abstract laws of appearance, the common source which is conducive to appearances and false illusions alike. And Hegel, too, considers the former to be of more importance and so he examines them in detail, while false illusions figure merely to the extent of a few paragraphs in the *Larger Logic.* Hence Goethe's theory of *'Urphänomen'* roughly approaches the view in which we use the 'phenomenon'-category today: it is a subjective husk in which the objective appearances manifest themselves. The objectivity of being for us and its subjective mediatedness form its structure simultaneously. It is an ambiguous configuration.

The 'phenomenon'-concept has manifested itself with a great many aspects in the history of philosophy. It may be part of different conceptions of the universe. But it has one common feature everywhere, namely that apart from the endeavours of the 20th century it points in the direction of the objective phenomenal. It is a category of the conceptual movement, proceeding from the illusions towards a more complete revelation of reality. This is how matters stood with Aristotle, and that is also the case with Goethe. This concept in which appearance and illusion are still in a mixed state is the unity of subjective distortions and objective elements, but it also has the increasing tendency to cast out the illusory and the contingent. Only the most recent development in philosophy has fetishized this dual aspect, determining the mobile and progressive character of this concept, taking away its particularity as a directional category and making it as proof of the inexplicability of the world.

The performance quality of the 'phenomenon'-category plainly shows that what we have been talking about here is not ontological presentation.

The reason, however, why we started the examination of presentation and illusion with this subject is that this is the primary configuration for human praxis: we encounter not appearances but 'phenomena', i.e. the presentational forms which praxis has changed spontaneously. But we cannot stop at this point, we must go on. There are two possibilities for us to choose from. We may either investigate the process of ontological presentation, the objective structure of the dialectic of appearance and essence, or trace the history of how the 'phenomenon'-form has become fetishized and petrified into an illusion. Our study will take a somewhat arbitrary course from here on. We propose to examine first the ontological structure of presentation so that in the light of this we can more distinctly separate the essentially epistemological structure of illusions.

This is undoubtedly a long jump but one which is required by the fact that transitions are lacking. But in the course of analyzing the difference between appearance and illusion we shall refer back to the continuity which leads in the development of human praxis through the 'making' of the false forms of appearance to appearances which are taken in an ontological sense. It is a matter of curiosity that phenomenology, particularly Husserl's late works, also attempts to transcend the sphere of appearances produced in the subject's performances and to develop a general theory of appearances. However, it was precisely the subjectivity of the performance quality which frustrated this attempt: the theoretical demolition of this subjectivity would have resulted in the differentiation between illusion and appearance, a differentiation which was clearly unacceptable to this theory. Hartmann has an excellent flair for observing the contradiction between the obvious possibilities and the limitation of the solutions. In the *Philosophie der Natur* [Berlin 1950, p. 39] Hartmann states that phenomenology has not fully carried out the conceptualization of the world in appearances because it concentrated solely on performance-'phenomena'; it overlooked the fact that there are also object-'phenomena' which have exactly the same form but do not manifest themselves merely in the observations but include in their contents every achievement of the positive sciences. But even Hartmann's measure is insufficient. What we are interested in is not the object-phenomena existing in science, but it is the presentational structure which is independent of man's presence (contribution). For indeed, my assumption is that the subjective element of reflection can be excluded from the phenomenal sphere only to a certain point, that the fraction of the relatedness to the subject will always be a constituent of objective phenomena. But we cannot prove the truth of this hypothesis at this stage. Perhaps this also accounts for our 'jump', away from that which manifests itself in praxis and into the ontological sphere. We are at the same time convinced that Marxism, once it has worked out its own theory of appearance (i.e. a phenomenology of its own), can successfully debate with and defeat the agnosticism of phenomenology if it starts out from this ontological base and examines thoroughly the various ways of how anything becomes a practical 'phenomenon'.

CHAPTER 2

THE GENESIS OF VIRTUAL REALITY

2.1. THE OBJECTIVITY OF ILLUSIONS

Hegel could think of illusions only as a transitional stage of development between cognition and the ontological modes of existence. He could not have a knowledge of petrified illusions functioning as reality even for prolonged periods and having an objective power for social practice and consciousness. The discovery of the relative objectivity of social illusions is due to Marx who with his fetish-theory created the basis for an overall description of the illusory structures and for the revelation of their laws. But the rise of objectivity and relative permanency made it necessary for a new contradiction to be resolved. For it is evident that, as compared with the objectivity of the social forms of presentation, illusions are of subjective origin. For one thing, illusions cannot be realized without a receiving and receptive consciousness, and for another — which is a different formulation of this same relationship between subject and object — the processes of nature, at least on an inorganic level, do not get started on the basis of illusory relationships. Hence illusion does not belong to the sphere of ontological objectivity. Still, it has a certain type of objectivity, for indeed illusion can be distinguished from error precisely in that the former manifests itself 'for everybody' in this 'being so'. And moreover, as an impenetrable outward configuration, illusion aggressively influences one 'from without' and therefore it is not solely the product of subjective ideas, fantasies etc. (Marx made numerous references to the fact that the presentational forms of capitalist society 'factually' assume some different configurations and that they become illusory not merely in the individuals' consciousness, interests and illusions.) For illusions to take effect there is a need for certain objective facts (or for the similarly objective covering up of the facts) even in the illusions of manipulation where the matter in question primarily concerns the changing of the 'way of seeing things', i.e. inducing blindness to real facts and readiness to accept spurious facts. How can this contradiction be set aright, namely so that the basically subjective nature of illusions, which ontologically depends on consciousness, be reconciled with their fundamentally objective, factual attributes, with characteristics of their presentational forms which have significance beyond individual consciousness and can be treated as though they were objective?

Our opinion is that the solution is furnished by the objective effect and ontological relation of 'being for others', a category which we scrutinized in the previous chapter with regard to social phenomena.

Illusions, like appearances, function on the basis of this category, but in the case of appearances this 'other' ('towards' which they work) may also be a material relationship of objective character, whereas with illusions this 'other' is only subjectivity, consciousness. In the case of illusions the sphere of presentation is determined solely by reflection. It is here that the particular objectivity of illusion arises, in that the socially guided performance of consciousness realizes the form of presentation into an illusion. The 'other' is not the abstract subject but it is the consciousness of the social individual, and hereby a false relationship is reflected in a medium in which the objectification is created by socialness. If we were facing simply an error, it would be only an individual confusion in reflection, without any possibility of objectification and therefore without a being for others. But illusions appear for 'everybody' like that so they have a certain 'being so' which is of conceptual origin and yet of objective validity. And in this respect, they are not purely subjective configurations any more: everybody must perform this subjective working of consciousness in the same way. Conceptual performance is socially guided. With reference to the illusion of wages Marx illustrates this reciprocal structure as follows: the forms of presentation manifest themselves with a certain degree of objectivity but will mature into illusions only as conceptual forms. "In respect to the phenomenal form, 'value and price of labour', or 'wages', as contrasted with the essential relation manifested therein, viz., the value and price of labour-power, the same difference holds that holds in respect to all phenomena and their hidden substratum. The former appear directly and spontaneously as current modes of thought; the latter must first be discovered by science." [K. Marx: *Capital,* transl. S. Moore and E. Aveling, Moscow 1977, Vol. I, p. 507.] This 'current mode of thought' to which Marx refers includes two important methodological references: first, illusions acquire their objectification in the performance of the thought, i.e., in the process of perception and interpretation. Until then they are only objective forms of presentation (phenomena). And secondly, as forms of thought they already have a secondary objectivity: they are produced not by individual consciousness but by the social exchange of ideas, by the conscious objectification of illusory relationships. That is why in society causal series can also be initiated on the basis of illusory relationships. It does not matter whether it is illusory or real appearances that have initiated a series of actions or a social movement. (The phenomena of fashion, for example, are collective stereotypes which, by veiling individual and special features and qualities, help to develop an illusory exterior and, as an obligatory form of conduct, a medium of false reality which consists of numerous individuals disguised as 'others', and where the illusory reaction produced must be accepted as reality even by the particular person who started it through his disguise.) This secondary objectification, the illusions as 'common forms of thought', will be dealt with later.

As a preliminary definition, we might say that the objectivity of the 'being so' of illusion exists merely for the subject, whereas the objectivity of appearances are guaranteed by their ontological effects.

Thus illusions arise only in an epistemological relation, in confrontation with cognitive consciousness. No illusion is produced without at least an elementary degree of consciousness, just as in inorganic nature processes do not begin on the basis of illusions. The fundamental condition of illusion is reflexive consciousness. The objectivity of illusions cannot arise without consciousness. That this paradox is resolved by the formalizing and generalizing medium of social consciousness we shall elaborate in our subsequent examinations. For now, we want to establish that the objectivity of the untruth of illusions cannot come into being without genuine and conscious subjectivity.

This means that the phenomenal elements manifesting themselves for consciousness become visible at all only from the point of view of a model of reflection (e.g. analogy, prejudice, ideological forms), and that their particular 'being so' will also develop in this reflection. But what is hereby 'seen' is of course not the essential relationship but a husk of subjective origin produced in the course of reflection. This husk does indeed carry in itself the fragments and the essential relationships of an objective appearance but it partly covers them and partly completes them with purely subjective notions, thus objectifying a false picture. Ontologically, provided that we consider our viewpoint independent of the observability on earth, there is an objective structure of presentation behind the illusion of the sun's orbit, namely that the stellar motion of the earth defined in relation to the sun will also present itself in the sun's relation to the earth. So the product is a certain transformed picture but it will become a true picture and also a false appearance only in our subjective reflection, insofar as we identify it with the objective and therefore not geometrically projected picture of the relationship between the earth and the sun. The principal feature of the 'subjectivity' of illusions is that certain objective phenomenal elements become visible at all only in the performance of a conceptual reinterpretation: this is where they assume their modes of existence (in fact their conceptual forms) and at the same time their falseness, too.

The peculiarity of illusions is that essence is 'being seen' in them without mediation. They explain themselves, or rather, they manifest themselves under the aegis of a 'false' essence. That is why Hegel says that "Show itself is, then, immediately determinate. Its content may be this or that; but whatever content it has is not posited by itself but belongs to it immediately." [Hegel: *Science of Logic, Op. cit.,* p. 22.] It is this contingent content that explains and, indeed, at all renders visible the concrete appearance as a false illusion. But where does this contingent content come from? It comes from the performance of reflection, from subjectivity. It is an analogous relationship in reflection, a relationship which has been taken from somewhere else and which renders the presentational form perceptible, realizable and into something that appears at all. This analogous attitude or analogical reasoning is one of the most important constitutive elements of illusions. Kant denied the objectivity of the sphere of appearances but clearly saw this subjective origin of the 'modes of existence' of illusions. Writing on transcendental illusions he states that this 'phenomenon'-sphere is originated solely in judgments and not in objects: "Still less justification have we for regarding appearance and illusion as being identical. For truth or illusion is not in the object, in so far as it is intuited but in the judgement about it, in so far as it is thought." [Kant: *Critique of Pure Reason,* transl. N. Kemp Smith, London 1929^{1}, 1973^{12}, p. 297.] He adds, moreover, that the senses cannot be wrong in themselves, so error belongs to the sphere of judgment. It follows from Kant's entire theory that for him this concept of illusion is a category of epistemology. It is subjective not only in its origin but also in its function. Kant cannot imagine illusions in which fragments of reality are also present, still less can he think of 'phenomena' which comprehend ontologically 'valid' and therefore empirically correct relationships — in an epistemologically false form. But he has a keen insight in realizing that the most important element of illusions is that they become objectified in the forming of judgments. The relatedness of things to the subject manifests itself and exists only in a form in which consciousness relates to the world of appearances, whereas the analogical solution gives the false presentational form of the originally objective 'phenomena'. It is this premiss which creates the immediate unity of appearance and (fictive) essence, i.e. illusion.

This outlook is the attitude of consciousness of immediacy, regarding as essentially the same those which are, on the basis of external marks, similar to one another. All illusions have this common characteristic. (The illusion of paid work, for example, is caused by the fact that labour is also a commodity. Its special feature separating it from all other commodities, namely that it brings into being more than its own value is lost sight of in this similarity.) But similarity, with respect to its origin, is again not merely a subjective category. It is a fact that labour-power has the character of a commodity just as it is a fact that commodity production changes all things, characteristics and values into commodities. The subjective element lies only in the erroneous generalization of the objective extrinsic similarity. And furthermore, we shall see not only that the analogue, the objectively existing 'similar entity' is a socially produced objective 'fact', but also that it determines peoples' subjective frame of mind. The development of averageness produces this objective 'similar entity' in terms of which we think and which, as the sphere of the choosable conditions, also determines our actions. The analogical approach is the form of consciousness of this average outward appearance of society, which blurs individual and at times essential differences.

This erroneous generalization occurs because analogical thought does not try to get access to the contents of appearances. In its view the problem of the structure of appearance has already been resolved because it already knows the structure which resembles it. In consequence of this it remains on the surface of appearances not only because this outer crust is impenetrable (and we shall return to the analysis of this matter in connection with our examination of the sphere of immediacy), but also because from the outset it concentrates not on this appearance but on the other one, the configuration known from somewhere else. The appearance under examination is only a reproduction of the one known from elsewhere and is therefore secondary. Analogy stops short before the speciality of appearances. What it can see in them is merely the frame known from somewhere else and in a different context. (That is why Hegel calls the analogical method merely a mental outlook and does not regard it as a conceptual or logical configuration. He sees it as a prerational formation.) This orientation towards something other (an orientation for which the special appearance occurring 'here' is merely an accidental instance of an already known relationship, the frame of mind of everyday consciousness without which it could not exist) is on which the quick judgment, the capacity to obtain a comprehensive view, the stabilization of knowledge etc. are based. However, as we are going to see, in modern society this frame of mind is fetishized and rigidifies into an infatuation which will itself prevent access to the things themselves. (The formal structure of fashion in dress, for example, is designed to appeal to this orientation towards something other and produces this way of seeing things. What is individual can be taken into consideration only as the analogue of the socially valid quality of form.)

The analogical way of thinking comes from an objective relationship (from the objectively existing outward similarity of things), and it is conducive to illusions only if it is absolutized and becomes an obstacle to discernment of the speciality of appearances, of distinctions. In this fetishized capacity, however, it is the principal means of producing illusions. It is at work from the simplest artificial illusions (counterfeiting, fraud, mimicry, hypocrisy) to social 'phenomena' of great importance (the illusion of 'paid work'). Manipulation also uses this analogical model when concealing the reality of dehumanization with the external conditions of a pleasant life, and manipulated consciousness accepts this mode of life, which to outward seeming is human, as reality.

But where are we to look for the element of objectivity, in this subjective postulatory performance? Can we talk at all about the objectivity of illusion if its mode of existence is created by consciousness? Let us reverse the question: can we imagine a subjective, i.e. fantastic or imaginary 'phenomenon' which is borne out in practice although it does not incorporate even the elements of ontological reality? The appropriate answer is provided by Hegel's genius, proposing that as an element or moment essence also lives in the falseness of illusion and so it is by undergoing a singular transformation that the true relationships are increasingly realized in the illusion. Though they are wrapped in a subjective cover, true relationships still assert themselves in it. In the *Larger Logic* he writes: "Two moments thus constitute Show: Nullity, which however persists, and Being, which however is Moment; or again negativity which is in itself and reflected immediacy. Consequently these moments are the moments of Essence itself: there is no Show of Being in Essence, nor Show of Essence in Being; Show in Essence is not the Show of another, but Show in itself, Show of Essence itself. Show is Essence itself in the determination of being." [Hegel: *Science of Logic, Op. cit.*, pp. 23—24. — Translator's note: Messrs Johnston and Struthers (as translators of Hegel's work) apparently used 'show' for what I have consistently used in this book as 'illusion'.] It is this brilliant insight which also illuminated the paradox of the history of science, namely that great errors had furthered its progress more than some of its important truths.

Essence in illusion manifests itself in a distorted shape. Consciousness concentrates mainly on the subjective presentational form with which essence is covered because it 'knows' this 'phenomenon'-form and not essence which is inherent in this form. Practice will nevertheless be able to grasp this element of reality within this form and to work accordingly. (This is the structure of the 'reality' of false consciousness.) Where shall we look for the origin of the hidden element of reality which is nevertheless contained in illusions?

We shall have to look for it in the dialectical relation of 'being for us'. We have already seen that the epistemological relation, the being of things for us and their practical and conscious appropriation constitute merely a sub-case of the ontological structure of 'being for others', which is the basic structure of being of appearances. The postulatory work of cognition can operate only with those phenomenal elements which manifest themselves for subject-being, i.e. in human praxis. In the primary, ontological sense things do not manifest themselves for consciousness but for subject-being, i.e. for the position which we occupy and from which we shall take in the world and proceed with our actions. And here already the laws of the phenomenal sphere apply: 'phenomena' will manifest themselves primarily for our situation in nature and society, for our existential data, and they will establish the possibilities for their different forms of manifestation with this objectively ontological interaction which has been formed with this condition of existence. The orbital movement of the sun does not manifest itself for the consciousness of the individual person. The phenomenon is the single-aspect transformation of the reciprocal relation of movement between two celestial bodies, the way in which this two-member relation manifests itself in projection on to one of the two members. This relation is given with respect to this very position, in its relation to the earth, and so this phenomenon is an inescapable and necessary 'phenomenon' for anyone who lives in this position. Whoever may come to the earth would see this phenomenon like that, even though from an extraterrestrial site, i.e. from a different position, he could observe the true relation of movement of the two celestial bodies and not the one which has been transformed on to one of them.

The 'position' from where the relationships look different is not merely a static point of view or a status in the social division of labour delineated by society. Only through the dynamics of action will it turn into a configuration which determines our outlook and preforms things. Appearances manifest themselves in one way for those who act collectively and in a different way for those who are 'excluded'. Joint activity, 'inherence', i.e. collective action is the medium which simultaneously preforms the effects tending in our direction and shapes the outlook through which we can see things.

Hence 'being for us' is preformed. Even before the reflexive work of consciousness can start, the natural and the social conditions of subject-being, the objective position of the individual or society, transform those appearances which are tending 'in our direction'. Their phenomenal forms are shaped by their being for our modes of existence and not by our consciousness which reflects only the 'phenomenon' shaped in this way either in its pure form as an appearance or in the false form of subjective hypothesis or an analogical explanation. The same transformation also takes effect in social practice. Before the individual capitalist can consciously gather information on the relationships of his practical world, the movement of total capital offers itself in a particular configuration of projection, as an objective phenomenon. And in the course of the process during which the total movement is pictured on the individual capital there will arise a transformed, projected appearance. From here, from this position, the total movement looks only like this, because it manifests itself for the individual capital objectively in this transformed configuration. And then, conscious recognition will try to identify this transformed picture with the total movement or to explain this presentational form by means of false analogies. But the primary projection of appearance is produced by ontological forces external to consciousness, consequently by the work of transformation. This will be an objective fragment of appearance, a fragment which includes the essential relation of the entire movement but in a distorted, projected form.

Thus, although it is true that there can be no illusion apart from the subject, without the postulatory work of reflection, nevertheless the elements of this presentational form will be preformed by the 'external' properties of subject-being, namely by the position that we occupy in nature and society. This total movement will be formed as a different objective projection in accordance with the position occupied in the division of labour at any given time. This projection-picture is of objective origin as an opportunity. It is subjective only in its realization, in the process of being formed into an appearance, and it becomes an interpreted picture, an illusion, only in consciousness. So when Hegel says that 'being of illusion is in another', namely in the differentiality of the conceptual explanation, then the ontological basis of this differentiality is accounted for by the fact that appearances do not manifest themselves for us, as for our consciousness but for us as for something else, directed towards our situation, our position in nature and in society. Whoever may get into our position would see the relationships only in this projection.

Here again, we meet the difference between natural and social phenomena. Natural phenomena also transform the overall movement of the totalities from whose contact they were born, and pass on this transformed, projected effect to their environment. Condensed into and projected on to these one or two correlations they make contact with things pertaining to their spheres of influence. But this transformation also carries to some degree the totality of the source of the appearance. Otherwise it could not exert its influence. (The level of radiation material and its frequency in the radiation spectrum both depend on the interaction between the irradiating substance, the screening effect of the media etc.) The situation is different in the case of social phenomena which exist in the order of conscious recognition and teleological postulation, so there is *ab ovo* a possibility that the transformed picture is taken for the totality of the relations since as an appearance it can exert its influence on conscious action in this manner, too. The reason why the transformed appearance can take effect in nature is because its impact is made not by its presentational form but by essential impulses. In social phenomena there arises the possibility that in conscious forms of action only the mediating husk, the presentational form, will take effect and not the influences which this husk mediates. (Although the motives of a lie form an integral part of the act of deceit, nevertheless only a lie which is believed to be truth will take effect. In the case of a well-played role it is only the role that makes its effect. The appearance will not impart its hidden substance.) The deliberately concealed motive is indeed transformed into an illusory action, but it is only this assumed aspect which will expand like ripple rings in social activities. In an ontological sense the lie is actually a form through which one tries to avoid the collision of certain situations and intentions and therefore it is a phenomenon. On this basis it is easy to distinguish the interactions which have produced the deliberate concealment, whereas the praxis of everyday life operates with directly recognizable signals, and this picture will not lead one to the real motives.

In social practice the simplest forms of deceit are of course the deliberately created illusions, which are easy to see through and are characteristic only of private relations. On the other hand, we regard as illusions those false forms of presentation which the overwhelming majority of society look upon as reality. But the same structure is at work even in this case. On the illusion of the interest on money lent, Marx writes that the illusion of 'money bringing forth money' is the result of an objective transformation. Between the two extremities of money-lending and returns with interest there comes into play the function it fulfills in industrial capital and the part it plays in the production of surplus value, thereby creating this 'surplus'. However, on the level of immediacy this process of transmission is lost sight of for the individual capitalist and it is objectively missing in the presentational forms. Only the money lent and then repaid with surplus figures in practice and only this is 'perceptible'. This presentational form is undoubtedly one of the projected pictures of the total movement of capital and, in this respect, it is also a phenomenon in the sense of ontology. But for the individual there exists only this form of presentation which, since it comprises the essential relations, does indeed in the abstract represent this relation but at the same time also hides it, allowing only the form of presentation to be seen. Consciousness, under the influence of earlier analogies, interprets only this visible picture. (Marx has recourse to the comparisons of circulation and trade as explanatory analogies which help to understand this phenomenon.)

Hence consciousness assumes a particular 'creative role' in the acceptance of illusions. It creates a virtual connection between the immediately appearing phenomenal elements since its thought is guided by the analogy of other relationships. This performance is facilitated by the formalism of everyday thought. In workaday thinking, a hotbed for the origination and the use of illusions, anything that, in some respects, manifests itself as formed seems at the same time substantial and essential. (The prejudices, the illusions of fashion, even the 'phenomena' of scientific progress, are expressed in such formulae, and their 'formedness' also serves as proof. The decimal system, for example, has given rise to the idea that the planets also have to conform to a decimal system.) This idea has often arisen in the course of the history of philosophy: with Aristotle's theory of forms and later with Leibniz and Descartes, formedness is a concomitant of substantiality, so every formula is at the same time a substantial relationship. Hartmann wittily refutes this ontological bias when he explains that the form is indifferent to relationships between the universal and the particular, the substantial and the insubstantial. [Cf. N. Hartmann: *Der Aufbau der realen Welt,* Berlin 1940, p. 238.] Nevertheless, this bias has survived not only in our workaday life but also in modern philosophy, in the phenomenological theory of forms and in formal ontology. We shall return later to an analysis of the structure of this formalism.

The formalism of everyday consciousness — with its spontaneous inclination to substantialize — facilitates our associating the forms of presentation as identical with the things themselves. This formalism will be the most important factor in the furthering of objectification. What is merely a chain of thought in perception becomes real when used as a formula.

In the foregoing we scrutinized only how the entirety of processes is transformed for some concrete social or natural position. But at the opposite pole of transformation a further category manifests itself, namely the functional form of presentation, a configuration attracted by the position. Position and function are two correlated categories, mutually forming one another. Position is an attribute of social role for which, irrespective of who fulfills it, people manifest themselves in certain anti-roles. It is only by means of their predetermined functional behaviour that they can approach it. The sergeant, standing on the rung of the military ladder, enjoys a position 'towards' which the recruits approach with a behaviour that is entirely different from that which they usually display. The only thing that the sergeant can see is that everybody confronting him is a 'stupid civilian' and so he identifies the picture manifesting itself in response to his position with the real person. On the other hand, the recruits adopt this attitude, this functional transformation, only with respect to the sergeant — without being aware of the fact. The same applies to the relation between confessor and confessant, official and applicant, boss and subordinate: position attracts a certain form of presentation, a characteristic configuration of a role, this being the way, and the only way, in which it can be approached. So the relationship between position and function — to quote Hegel — is a 'reflexive determinateness'. The one exists only in relation to the other. Marx has a witty comment on this category in *Capital,* saying that such reflexive determinatenesses are anyway peculiar things, as some man is, for example, king only because in relation to him people are his subjects, whereas these think just the other way round, namely that they are subjects because that one is king. The correlation, just as the entire process of the transposal, is of an ontological structure, but in its simplest form it manifests itself as a conscious adjustment, as the functionality of a role that has been assumed. The mode of conduct which we must adopt for the sake of successful activities is shaped by the expectations of certain social 'positions'. Within this expectation we may opt for various alternatives, but we can attain a certain optimum value of efficiency only through some specific forms of behaviour. (The pupil under the guidance of a poor teacher will give the kind of answer that he feels is expected of him, that is, he pretends to be a good sort, for this is the right behaviour in response to the teacher's role. A similar logic is found — on a more dangerous scale — in careerism, obsequiousness and in the monolithic way of thinking formed during the dogmatic period of a movement.)

Thus, these two different kinds of determinatenesses are an objective interaction. To this we must add that both categories are institutionalized modes of existence in the social division of labour. Every 'position' forms a link in the institutionally developing total social praxis not only as a stepping stone in the official career structure but also as a situation, as a relatively broad or restricted space for movement which was occupied or staked out even before the appearing of the individual in production, intellectual life, and family and social relations. It becomes a reflexive determinateness only when two coherent configurations are examined with regard to their reciprocal influence. But we also know that what we examine as a function in its attitude towards position is also a position in such a relationship of social connections as directed towards something else and that what we can see as a position means at the same time an aim, a 'being for others'. Society, as an organism, is based on these many different kinds of intertwining structures. Both these categories are ontological data whose interdependence and content, as well as the course of action taken by those living in them, are determined by the total movement of society. In other words, position determines the primariness of being as an 'overall' element in this mutually developed space for movement.

'Being for others' is thus the mode of existence of position. It is a more concealed configuration which presents itself rarely without mediation, whereas its effects, the way in which it exists for others and how others 'respond' to its signals, form the space of immediate practice for movement. Everyone's action has its origin in an *a priori* given position, and the objective effect of his starting-point will persist in his actions. (It is like the prank played by children when they put a tag on someone's back with the inscription 'don't anyone talk to me', and the passive, unsuspecting party wonders why he receives no answer and why he just gets laughed at, regardless of what he does.) The relation to position is of course not so unselfconscious. Its requirements and effects may become palpable in the form of false consciousness or in class-consciousness. For all that, the interactions of immediate practice and the progressively emancipated objective medium of its exchange of activities are furnished by the difference between the objective effect of 'being for others' and the subjective, teleological directedness as well as by the combination of these two.

However, there is an important difference between these two 'basic roles', namely that the function is active, whereas positional existence is mostly passive; and conversely, the passive formedness of the function remains imperceptible to man, while the actively effective and challenging power of positional existence also remains invisible. The epistemological and ontological spheres are thus separated in this reflexive relationship and so make room between them for illusions. Let us translate the abstract formula: Functional existence, i.e., the mode of manifestation 'for others', is a directedness, a 'being' which is determined from without and 'called' from the direction of position, though teleologically we steer it from within. But it is precisely our needs and interests defined in our goals which compel us to adopt a well-defined role of approximate accuracy or a form of activity in approaching those positions which determine our action. However conscious I may be of the teleological determinateness of this attitude and this series of activities, namely that whatever happens I must attain my goals, I am quite in the dark about the fact that it is not I who decide even my very first steps but that they are preformed from without by one of the socially given positions confronting me. That is to say, this position surrounds me with a space for opportunities, directing my behaviour in advance. What I can survey is merely the orderly state of my goals and my means employed, but I cannot see that the position to which I am attached has formed in advance the range of my thoughts and ideas, i.e., my needs and consequently my goals, and that it determines my conduct when this involves a series of actions aimed towards these goals. Though teleologically this is all 'mine', in the end I am steered also from 'without'. My goals, steps and possibilities are preformed. And this also means that in this 'functional existence', in the directedness 'for other', this teleological illusion, namely that I pursue my own goals, conceals the genuine social aspect of position. I can see only as much as it turns towards me, precisely as much as my directedness allows me to see. I am surrounded by the forms of self-deception. Whilst functional existence is an active pursuit which is aimed at something, positional existence enters consciousness as a receptive attitude, as a passive existence. The person 'confined' within this position cannot see his own activity, his provoking role which 'preforms' others and which is radiated to others from his situation, from his objective positional existence and which determines *a priori* the attitudes directed towards it. The teacher can see only that he is surrounded by obedient, even over-zealous children, and he puts this down for his successful method of teaching, notwithstanding that it is a preforming effect consequent upon his position (perhaps upon the 'Prussian rigour' of the school).

The objectivity of social situations exerts its influence by preforming attitudes and appearances and forcing them to assume a certain 'controlled existence', a controlled form of presentation. The most important version of this 'controlled' form is not conscious play-acting, behaviour, demeanour or the consciously formed medium of manipulation but the structure of the objective connections of social phenomena. Here, too, the presentational forms take on the shape of their 'being so' for a certain conception, while they will be reacted upon with a creative force by whatever they exist for. It is in this sense that Marx writes, for example, on the dialectic between the forms of value in use and those of exchange value. Both categories are phenomena of the objective economic sphere but they nevertheless relate to one another with a certain 'directedness'. He writes in the *Grundrisse* that circulation, as the realization of exchange value, implies (1) that my product is a product only if it is a product for somebody else; so it is a specialty discontinued to become a generality, and (2) that this product exists for me only if it has already been appropriated for somebody else and so it has come to exist for somebody else. Thus, the exchange process creates objective categories of reciprocity in which certain functional configurations assume their form-determination only for a position with which they are confronted. Of similar structure is the relation between commodity and money, labour and capital. The form of 'being for others' is determined by the material counterpole of the relation since the position 'towards' which it exists will materialize.

And at this point we can see a further distinction between the social and the natural forms of presentation. Natural phenomena are not 'controlled', their ontological phenomenal form does not depend on that for which they exist. (It does not matter for the sun's radiation whether it quickens the flora of the earth or affects people's spirit.) In society, however, the particular quality of functional existence is determined by the position towards which it is directed. The aim of its directedness is creative and it reacts on it with a creative force.

It is obvious that this functional existence already carries in itself the element of illusion, for indeed it is easy to regard the traits of role-playing as the primary data of things. All fetishism is based on this, from the fetish-like nature of money to the alienated illusions originating in the personality cult. And furthermore, these illusions are created by the concerted action of function and position. The unreal plans coming from the top are complemented by the likewise unreal reports which are, however, believed to be a must for position: we regard the charismatic wisdom and the infallibility of the leader as his personal characteristics and not as a false element of consciousness which we have also helped to create. And still, we must say that this functional existence is essentially not yet an illusion but that it is an objective, ontological type of appearance. It becomes an illusion only in the epistemological relation when we cannot survey the other pole of the relational existence and in consequence we cannot see the secondary objectivity of the appearance either, an objectivity which presents itself for our position.

What do we mean by the foregoing? The answer runs like this: the relations which determine social activities are mediated by things which in their mediating capacity gain the same kind of objectivity as is possessed by their natural characteristics. But this second nature can be understood only from this capacity in which they mediate social activity. Exchange value, for example, is just as much an objective category as the physical property of gold is: money, as legal tender or as a unit of measurement, is not an illusory property of this metal but it is a phenomenal concretization of a social relation, notwithstanding that this phenomenal existence 'sits' as a secondary one on its original physical properties and is interwoven with them. The objectivity of its secondary nature will be revealed only when we disregard its characteristic feature, namely that we are immediately doomed to failure in social praxis just as the romantic and utopian notions which rebelled against the function of gold as money but which wanted to justify this attack by demonstrating the natural property of gold were from the outset doomed to failure. Functional existence results from the objective concatenation of the division of labour, from the total social practice of economic relations, and within this range — in the dialectic of position and function — it dictates man's active approach to things with an objective, inescapable 'being so' just as any other natural processes do. However, consciousness will be able to perceive and understand its phenomenal nature and its essential laws, or according to Marx, the concrescence of natural and social characteristics, only when it can see the dialectical relations of the interactions between 'being for others' and the 'other', the concerted action of function and position. If consciousness can see only functional existence, and the relation remains inaccessible within which the 'mode of being for others' arises, it will inevitably be unable to distinguish clearly what pertains to the natural structure of things and what has been built upon it as a social relation. The phenomenal nature of functional existence becomes an illusion. (If the relation which measures the work of society as a whole by its individual form of realization, i.e., the reciprocal relation of the product of exchange value and use value, cannot be seen, then gold, as the medium of universal exchange value, will, in its natural capacity, become money, and the illusion will arise that it possesses this property *de natura*.)

This functional existence is the basis of bureaucratization in everyday life: the order of the power structure becomes indistinct, and official function and human characteristics are interwoven. 'From without' it seems as if the 'local value' of the desk were the personal and impersonal characteristic of the individual people. Kafka, in his satire on institutions, puts this grotesque identification and illusion in the centre of his method of description. As Günther Anders writes: "Many of the characters are nothing else but mere functions. One man is only a messenger and nothing else; one woman is only 'proper connections' and nothing else... And this 'nothing else' is not even an invention of Kafka's but it is a creation of modern reality in which one can exist only in the role of a special function: one is only a minute particle of one's post." [Verbatim translation. Anders: *Kafka, pro und contra,* München 1951, pp. 46—47.] The bureaucratic relation does indeed obscure the social division of labour, the ultimate meaning of the hierarchy of the engaging cogwheels, and that is why, to outward seeming, human existence becomes physically, too, identical with social function. In Kafka's world the people become eerie chimeras because the hierarchy of power whose irradiations produce these roles remains invisible and only the unmediated function is seen. The same applies to the illusions of the personality cult: the commanding charisma stands for its own sake, and we cannot see the order of manipulation which produces this relational existence. We cannot see the fact that the production and the political forces of society as a whole are projected on the leader's position and that by this means his human personality is amplified by the addition of a social power.

Two things must be said in connection with the role-playing forms of social conduct, over and above the fact that they produce illusions. For one thing, modern sociology tends not only to overrate but also to make a fetish of the category of objectivity of the role. Jürgen Habermas aptly points out that it is only at an advanced stage of historical development that there appears a degree of institutional relations in which interlinked functions already have a certain objective 'framework', but that even here it is the concreteness of the alternative conduct of the individuals and of historical missions that turns social status into a genuine functionary role. [Cf. J. Habermas: *Theorie und Praxis,* Neuwied/R. 1963, p. 173 ff.] A second consequence of this is that certain functional relations represent the 'skeleton' of social existence and therefore, however much we feel its alienating influence and rebel against its fetishizing role, this configuration can never be given a fully subjective character. Consequently, depersonalization and functionalization are indeed products of a certain alienating process but some of their elements will remain even if the total phenomenon of alienation disappears. We cannot be present with our entire individuality in all our actions and contacts (functionalism also has an exonerating role, allowing of quick reactions and of actions pursued simultaneously in several directions), and through the increasing complexity of social relations this process is partly reduced and partly strengthened. When we are looking for one of the sources of illusions in functional existence we must seek for its reconciliation not in the utopia of a society 'without roles' but in the development of a clearly arranged world and a consciousness that is free from any kind of fetish. Thinking in terms of functions or roles will produce illusions only if the role assumes an independent form and becomes a quasi-reality by means of turning against the personal characteristics of its performer and not only conceals him but also replaces his essence with something else.

So an illusion arises when the relation in which it is distinctly separated disappears, and the mode of 'being for others' becomes visibly connected with that 'other' which attracts and produces this phenomenal existence. The 'secondary objectivity' of social relations becomes identical with the primary, natural characteristics of things and people, because the social position which elicits their particular social abilities cannot be seen. Social illusions are thus created as a result of a twofold process. On the one hand, they conceal the poles of the connections of relational existence, allowing only one extreme, i.e. role-playing existence, to be seen. And on the other hand, its irrational quality which is incomprehensible in itself and in its natural characteristics must still be accounted for; this work is done by subjective interpretation: the missing members of the mediations are complemented by a false intellectual mediation. And what this socially forming attitude adds to the originally functional form of presentation becomes at the same time objectified: it succeeds the invisible objective mediations and assumes their sphere of action.

Teleological praxis and the process of the attendant scientific cognition endeavour to strip away this anthropomorphic outlook and to grasp things in their objective relations. The way this happens is that the data appearing are conceived of as the relational existence of ever new ('phenomenon'-like) totalities which is time and again queried by widening practice, while science which is driven by these stimulating effects reveals ever more accurate — or new — relations of totality. So in cognition, in transcending of the apparential form 'for us', we are confronted not with the functional form but with the characteristic transformation of the total movement. But before even the broad outlines of this total movement can be sketched, a long-term development of science is required. In other words, consciousness progresses by the aid of 'false totalizations'. The conceptual form of the approximating nature of cognition is the positing, refining, rejecting and replacing of ever new virtual totalities. Karel Kosik uses the concept of 'false totality' only in the negative sense, as the category of false reality, of manipulation. [Cf. K. Kosik: *Die Dialektik des Konkreten,* Frankfurt/M. 1967, 1971[2].] Indeed, neither human cognition nor praxis would be conceivable without the successive, ever improving series of false totalizations. The reason why they would not be conceivable is that praxis, and with it the point of departure for teleologically steered cognition, is always the individual appearance and its projectional and transformed totality image. The sphere of life in which we work and live is ruled only in the abstract by full totalities overlapping each other, and indeed it is the 'being in each other' that creates this transformed presentational form, the objective multitude of transposals, and only this manifoldly pictured and transformed configuration reaches the individual person; so the original overlapping totalities will have to be reconstructed on this basis. The total movement therefore is inherent in the (perhaps functional) appearances which manifest themselves to individuals, but we can see neither (1) the metamorphosis that this movement has suffered in the course of the mediations nor (2) the noumenality that the totality possesses. And finally, as regards what we have hitherto taken into consideration as the decisive element of the 'blindness' of unmediated praxis, we must first separate what manifests itself only for our subject-being, for our individuality, from the socially general and average form of presentation. An individual's consciousness proceeds from pictures on two projections: from the objective and transpositional form of totalities (i.e. from the immediacy of 'presentational forms') and from their 'special' 'phenomenal' nature which manifests itself for his position and subject-being. Only by transcending this double projection can consciousness reach genuine totality. For indeed, if 'only the whole is true' (Hegel), then it will be inevitable that cognition will be able to orient itself only in 'false', inexactly assessed totalities which are always blended with subjective elements. Cognition, as a matter of fact, performs an extrapolation: it objectivizes the principles of motion of the praxis-controlled piece of reality on to the whole. At various historical stages of cognition and in different proportions this extrapolation embraces the real essential relationship and false and subjective 'complementation', i.e. virtual totalization.

So essential appearance, i.e. relational existence, is reproduced by thought and by science. Yet the resulting totalities are not simply projections of errors but the approximations of genuine totalities and their subjective and conceptual complements. In the progress of science both objective relationships and imaginary, hypothetical interpretations are objectivized by successive and complementary theoretical systems. John Selye describes, for example, how the Wassermann reaction represented essentially a false point of departure, a false theoretical totalization, in the discovery of the pathogens of syphilis. In fact, the preparation (liver extract from stillborn children affected with syphilis) was merely a specific, accidental form of presentation of the more general phenomenon. It later became clear that an antigen of the same potency can be obtained not only from a healthy liver but from other healthy organs as well. So, while the test grasped a genuine relationship between the phenomenon observed and its totalization, it at the same time hypothesized a false totality. It 'complemented' the phenomenon in the wrong direction. And still — writes Selye — this false theory has proved more fertile than thousands of correct ones. [Cf. his *From Dream to Discovery: On Being a Scientist.*]

Social illusions are also produced in terms of a similar logical sequence; the relation between the transformation of appearances and their 'own totalities' is not visible in its entirety. Consciousness, however, increasingly reveals the hidden connection, but in doing so it not only happens upon the original totality but also complements it with its subjective ideas and with analogies, subsuming whatever has been brought to light in what are only theoretical totalities. Illusion is both the unity and the objectification of this twofold operation.

Accordingly, false totalization is an objectification whose reality is a limited sphere of appearance, and its falsehood is the subjective amplification of this limited sphere of validity. Indeed, if we do not notice that the given 'phenomenon' is influenced also by greater totalities which are beyond the known sphere of practice, we shall obtain illusions.

We identify the deficient phenomenal relationship with the law of the greater totality. The monetary system set out from the most tangible unit of the circulation of money and described its relevancies in its laws. The reason why the operation was a false totalization is that it did not make allowance for the fact that the circulation of money is merely part of a larger unit and that it is simultaneously its constituent: it is subject to the production process, while it also produces its 'tracks'. What the monetary system described was for the most part true (namely the first systematic survey of the currency of money), but the way in which this law was postulated as the essence of the total movement created by the same token a virtual essence. This historical-ideological form was real not only as far as its partial results were concerned but with its false totalization it played a genuine social role as well: it helped to eliminate usurers' capital, the 'antediluvian' forms of capital (Marx), to make capitalist circulation of money universal, to break through feudal restrictedness. (In connection with the analysis of false consciousness and ideology we shall later return to the subject concerned with the social and practical 'truth' of false conceptual forms.)

Hence what is decisive in the difference between the illusory and the real is the objective order of magnitude and the subjective reference. Consciousness extends the recognized relationship to a larger unit, regarding this extrapolated relationship as essence. Here the ontological structure is as follows: the 'limited sphere' from which the science of a period sets out is an objective datum itself. Its frames of reference are not created by the arbitrariness of subjectivity but they are determined by the radius of action of socio-historical praxis, circumscribing the scope of the survey; this objective and practical boundary will not allow the tracing of relations extending beyond it. The monetary system and the resulting illusion — namely that the circulation of money is the model of the total movement of production capital — did not come about by virtue of the phenomenal sphere being arbitrarily narrowed down, but followed from the historical and practical fact that since not all branches of production had been subordinated to industrial capital, monetary funds were the prevailing form of capital for the time being. The capitalistic process of production was then indeed a subordinate part of the circulation of money, and the sphere of the real totality was factually, too, more restricted. The comprehensible sphere of appearances was limited in its objective nature. That is why the consciousness of that time was compelled by a certain objective necessity to regard this totality as 'final', i.e. essentially determinant. Hence the apprehensible extension of the relations of the phenomenal sphere depends on the historical conditions of human praxis, and if this sphere happens to be 'limited' then there is a certain kind of objectivity in making it universal and in terms of virtuality 'essential'. It is not an arbitrary objectivization but it is the result of the restrictedness of the historical age.

Let us summarize now the, for the time being, elementary and conceptually most primitive conditions of illusions. Illusions manifest themselves only for a certain position: the virtual motion of the sun can be seen only from the earth; only a man deeply in love with his beloved will not see her faults... Illusions occur in the immediacy of 'being for us'. But what also follows from their existing for a certain position — whoever may be in this position — is that this virtual picture is also objective: anyone who comes into this position and views the appearance from here will find himself confronted with this virtual configuration without, of course, acquiring any knowledge of this fact. This is where we must seek the philosophical bases for the objectivity of illusions. But no illusion will be produced merely by a false relationship manifesting itself for a certain position *a priori,* i.e. without subjective participation. For the other essential constituent of the illusion is subjectivity, the participation of reflexive consciousness. There is no illusion without consciousness and, in contrast to social processes, no natural processes are initiated on the basis of illusions. So the other elementary constituent of illusion is the activity of consciousness, namely analogical explanation and interpretation. Whatever presents itself for position will be analyzed by consciousness which will more or less automatically (spontaneously) classify the image together with similar ones which are already known. This is how consciousness 'totalizes': on the model of what is already known it interprets what is still unknown or only formally known. So it creates the illusion on the basis of analogy. In this respect, illusion is indeed a 'created nature'. These two components (the position for which illusion presents itself, and the analogical thinking which gives it the appearance of truth, makes it existent, i.e. creates it) are the most elementary factors in constituting an illusion. The operation of both of them, however, is alike conditional on there being an immediate connection between the subject (consciousness) and the object (illusory picture) and no possibility of attaining the appearance by some other mediated approach. (A lie is always regarded as truth until we have convinced ourselves by other means that the facts are otherwise.) These two creators of illusion thus operate only in the sphere of immediacy. This is what we have to examine now in detail.

2.2. IMMEDIACY AND THE CONDITIONS ALLOWING THE OCCURRENCE OF ILLUSIONS

Illusions occur in immediacy. We have already touched on this category several times but only in one sense, the epistemological one: 'being for us' becomes an illusion only if for some reason or other cognition is unable to take in the structure of the components of the appearances and their hidden background and it remains within the sphere of the particular and 'direct' experiences instead.[1] However, immediacy is a category with a twofold aspect: besides its epistemological form it also has an ontological one. And moreover, the immediacy of cognition is built in the last analysis on the objective 'process of immediatization' of things, and the empirical, cognitive and applicable relations of everyday life come from the entwining of these two dimensions.

First of all, let us sum up the subjective element of immediacy. Every cognitive process proceeds from an immediate relation with reality. Whether we are associated with things which manifest themselves in praxis or with the traditional and familiar precepts, the first step of acceptance is always immediate. Perception, the use of instruments, emotional and spontaneous reactions or the responses of the particularity of life form the point of departure for all degrees of cognition which are of a higher order and therefore mediate.

Moreover, by virtue of its emotional and impulsive characteristics, immediacy emerges mostly as an enticement by which it entraps a man. The gratification of his desires and immediate interests diverts him from his basic human requirements and influences his outlook or even his entire disposition. This enticement is fraught with peril, particularly in the modern, manipulated way of life, in a gratification which begets frustration. One opts for alternatives which, though they are pleasant enough and satisfy one for a while, in the long run will fill one with discontent and an inexplicable source of being out of sorts. The pitfall of the immediacy of thought is that emotional spontaneity becomes fetishized.

[1] Subjective idealism and through it agnosticism, a category which was given general currency especially by Kant. According to Berkeley's thesis, perceived and existent things are identical and degrees of cognition are by the same token degrees of reality of things. "The very being of a sensation or idea consists in being perceived." [*The Principles of Human Knowledge,* 1710, Paragraph 90.] He forgets to realize that illusions are also 'perceived' and considered real: differentiation could begin only after this perception.

On the other hand, it is undeniable that our immediate relation with our environment is an unavoidable starting-point and that at the same time it amounts to more than a mere beginning. That is why we have emphasized that this relation can be only the first step towards reality since even the category of immediacy taken in an epistemological sense is more than an epistemological relation.

It is preceded by practice in which the first mediations are created and cognition attaches itself to this partial mediation as an immediacy. That is why we say that the immediate appearances for consciousness are at the same time 'more' than only immediate data. No cognitive process would be possible if we wanted to orientate ourselves only in immediacy, making contact with a single aspect of things. Consider, for example, the difference between an experience (adventure) and experience (expertise). The former represents the reign of the immediacy of the world over our emotions and our thoughts: it captivates us by suddenly breaking in on us and dominating over us. Situations and objects surround us, dictating the picture formation of reflection only with the aspect with which they 'surprised' us. We cannot step out of its circle. We accept what breaks in on us. In this sense immediacy is an event of elemental power within which some abstract features coexist without differentiation with the naiveté and inexperience of our emotional reactions. An experience is the model of immediacy taken in a subjective sense: it adds something to the one-sidedness of appearances and what is 'given' in it is only that 'which' presents itself and 'as' it presents itself. Whereas experience — which is usually also regarded as something that pertains to immediacy — already breaks out of this subjective and emotional restrictedness. It uses the 'cunning of reason' because it 'experiments' with the manifold characteristics of things. It wants, too, to discover their hidden properties though they can be approached only empirically, and to this extent it is already an ancillary concept of the objective expansivity of practice and not merely of the one-sidedness of consciousness and cognition. The *savoir-vivre* of the 'experienced man' exceeds the immediately visible errors which are derived from one-sided information; he also has a knowledge of the unfamiliar. In other words, even the subjective concept of immediacy can work only by transcending the stages of the mere function of consciousness (perception—experience—deliberation). Nevertheless, this type belongs to the subjective pole of immediacy.

That is why in our quest for the source of the objectivity of illusions the other meaning of immediacy is more important to us, namely 'immediatization' taken in an ontological sense. It represents the mode of presentation in which the complex relationships and motions manifest themselves in their transformation on to only a single presentational form and so not only for consciousness but also for the material world. The concept no longer implies that individual consciousness is restricted. The reason for immediacy is not the simplicity of thought and experience but an objectively (socially) produced configuration. A man's character is shaped by his biological endowments, his education, his living conditions and his lucky or luckless fate, but in a decisive encounter this prior history, the age of mediatedness, will disappear in the simple and direct form of the gesture and only the immediate decision will take effect and be visible. The 'mediatedness' of the individual is condensed into a single gesture and it thus becomes immediate.

Hegel had a clear view of this ambiguity, of the cognitive and ontological characteristic. Hence with him reflection and immediacy form a concept. "Show is this immediate Non-existence, and it is this in the determinateness of Being. There remains for it therefore only the pure determinateness of immediacy; it exists as a reflected immediacy, that is, as that immediacy which exists only through the mediation of its negation and is nothing in relation to its mediation but the empty determination of the immediacy of Non-existence." [Hegel: *Science of Logic, Op. cit.*, Vol. 2, p. 22.] So the term 'reflected immediacy' has two meanings. On the one hand, it refers to the restrictedness of cognition, to our attitude in being inclined in our particularity to regard things as they manifest themselves, and do not search for the 'back' layers. But on the other hand, this reflexivity also means that even the things themselves manifest themselves through a single 'channel' only and that a sole objective medium allows them to take effect and come to the front. And this single means of approach now depends on neither the laziness nor the limitation of the subject but it is an objective datum, a framework which we can only gradually transcend by the roundabout means of scientific development. A crystalline structure or the composition of the protein molecule is simply 'given like that'. It conceals the sequence of its development, of its determinative mediatedness with its 'being so'. What has evolved in the refining process of thousands of years presents itself objectively, too, as a simple immediate fact for praxis and cognition. The mediatedness of things has turned into the immediacy of their mode of existence.

It is to be remarked only parenthetically that while this immediacy is capable of being penetrated through scientific means, by the conceptual and practical machinations of mankind, it can never be eliminated completely. The reason for this is not only because in the course of our work we tend to create ever new relations of immediacy (our relations to the means we employ is always immediate) but chiefly because our anthropological data (from our sensory organs to our reflection) will always have a share in cognition beyond a certain limit. A fully adequate natural picture simply does not exist. And while in social orientation the scientific discoveries do destroy illusions produced by immediacy, everyday life on the other hand proceeds a great deal more slowly in espousing these achievements and can orientate itself more comfortably in the circle of immediate data. And moreover, this is even encouraged by the fetishizing processes of modern capitalism, chaining the particular man to the configurations which manifest themselves immediately by virtue of his emotional reactions, his base or simple impulses or his love of comfort. (The capitalist can make a better use of the alienated and false illusion of profit than of its scientific description: the working class was for a long time deceived by wages and later by the illusion of receiving 'shares', albeit the theory of surplus value has been intellectual common property for a hundred years already.) The 'single-channel way of approach', the one-sided manifestation of essential processes, is here the result of the objective (and artificially intensified) process of a social system trying to conceal things.

As immediacy is capable of being penetrated only in a relative sense, no absolute immediacy can be created either; all our contacts with the outside world are realized through different media. That is the basis of human civilization: the instruments of production, social situation, language, status, parentage etc. are all media which connect us with reality in the course of a series of transposals. And these media even proliferate with the development of civilization. A man's everyday life unfolds between two poles. His immediate relation to things is assisted by a series of familiar and therefore already invisible media. Nevertheless it would be absurd to confine a man to the sphere of mediacy. Spontaneity, particularity and direct contact with things and people are elementary conditions of our life. One could not possibly live without the corrections, the emotional and intellectual inspirations, the pleasure or the sufferings that are all inherent in this sphere. The sterility of the way of life of the intelligentsia results largely from the illusion that it is possible to achieve a life-style which lacks all particularity and simply 'leaves out' immediacy. Everyday life routine which is worthy of man means only that we can ensure a free flow between emotional and conscious reception and elaboration of the immediate effects of the world (i.e. 'open to reality') and the survey and creation of the systems of mediacy.

All this points to the fundamental ontological law of immediacy, a law which Hegel set down first: immediacy is the mode of existence of mediations: relationships which bring forth and mediate individual life can exist and produce effects only in an unmediated way.[2] Behind the 'fact' of the extinction of a species of animals there are a thousand webs of mediations, but the result will carry these causative transposals only concealed in the husk of the mere fact. The death of a particular animal or its inability to breed is an 'individual' case which can also be accounted for by some sort of an 'accidental' misfortune. What produces an effect, what presents itself, is an immediate fact which embodies and changes into an immediate datum that which is the result of a long process and a complex constellation of events. This general ontological relationship, the process of 'immediatization', even expands in society. Here the conceptual mediations also contribute to the creation of a gesture or an appearance. The occurrence of tendencies is mediated not only by many different kinds of inducements but it is also affected by intellectual motives which, so far as the result is concerned, exert almost as substantial an influence as the material incentives. But even if these ideological mainsprings indeed disappear in the 'downright facts', they can perhaps be found again there more easily than the material ones. The process of immediatization conceals itself, only to betray its emergence by misleading hints. Thus it is here that the opportunity for illusion manifests itself. Ontologically, as regards the total social movement, all individual activities are simultaneously controlled and controlling, social and individual. For conscious practice, i.e. on the subjective side, this duality is incomprehensible. The individual person cannot fathom what sort of mediations, transpositional effects coming from afar are influencing his desires and his intentions, or changing the results of his actions. His particularity will conceal the mediatedness of his existence, so he will orientate himself in a medium which is partly real and partly illusory.

[2] "There is nothing in Heaven, Nature, Spirit, or anywhere else, which does not contain immediacy as well as mediacy, so that these two determinations are seen to be unseparated and inseparable, and the opposition between them is null." [Hegel: *Science of Logic, Op. cit.,* Vol. 1, p. 80.]

Immediacy is therefore the mode of existence of mediations. It is produced not only by the restrictedness of consciousness. Reality itself also narrows down its presentational configuration. Let us see by means of an example how this conformity with a natural law takes effect: I want to get off the bus but I forget to signal in due time. The driver picks a quarrel with me, hurling insults in my face that far exceed the measure of the mistake I have made. Our squabble seems to be an immediate affair between just the two of us: the bus-driver is angry with me and, on the face of it, with good reason. But this is only an illusion, as the story preparatory to the quarrel, the process which explodes precisely in my absent-mindedness, has an objective share in this scene. For indeed, it is not me whom the driver wants to punish. It is not my person with whom he is annoyed but with the series of other absent-minded persons who have on countless occasions made the same mistake before me, with the passengers who had already made him lose his temper even before I got on the bus. The absent-mindedness of the others — as a medium — is between the two of us. It is, however, not seen since it occurred earlier, and I only happened to assume the kind of role which my predecessors have created and for which the driver had already prepared his obviously righteous burst of anger. The immediate relationship, i.e., the quarrel, conceals the antecedents, the mediatedness of the scene. This is the phenomenon of 'immediatization' taken in an objective sense, and compared with this my immediate reaction, i.e., my subjective and conscious attitude of immediacy, can only be secondary. I am simply blinded by moments of time which compel me to take decisions and to act as well as by my own passionate defence of my interests. I am unable to grasp the perhaps guessable background of the scene. The appearance itself has turned into an immediate one, while the emotional and conceptual restrictedness of my reactions is merely the subjective amplification of this objective 'immediatization'.

By generalizing this example we can draw several consequences. First of all, the mediated relation occurs at the intersection of individual actions and social praxis. The individual simultaneously acts as a subject, as the initiator of his seemingly self-evident actions, and as the predicate of an earlier process of mediation, as one of the links in this chain. His actions and his accomplishments are simultaneously derived from him and from the intangible, impersonal and socially given 'other'. Immediate practice conceals the threads of the transposals. So for the individual the aspect assumed by his activity and the result thereof is different from the way this very same result would look on the level of total social praxis.[3]

The other consequence is that immediate relation works as a singularly irresistible husk. None of us can step out of it. It determines our actions, as it were, 'from without'. Our responses are given in advance because we create the scope of movement for one another. In fact of course we merely pass on the effects which assert their power through us and invisibly to us. This immediate relation which encloses the two of us together from without indicates that immediacy is a category of totality. The threads that move us are invisible only 'within a given circle', while in a greater totality they will already reveal their hidden relationships. But the problem is that it is almost impossible to get outside of the given totality. The individual person is always 'within' the magic circle drawn by the situations and he cannot obtain the clear view of an 'outside observer'. This totality is the unity of subject and object in practice, a connection in which one grasps his object in such a way that in the end the object does not let one go. This internal, active relation helps to build up the 'external' bounding husk as well.

This is a commonly known sociological fact. Albeit in a superficial and reduced form, and recognizing neither the implications resulting from social praxis nor the identity inherent in the difference, Vierkant formulates this relationship as follows: The active individual is always a prisoner of his situation. He is not only biassed but also limited in his ability to take in the situation, whereas the observer's judgment is more lucid and his capacity for maintaining standards is correspondingly greater. [Cf. Alfred Vierkant: *Gesellschaftslehre,* Stuttgart 1923, p. 392 ff.] The idea can be used in its germ, though Vierkant does not realize that even the most contemplative outlook is an active participation, whereas action is a contemplation of 'passing judgment' not only in an identical context but also with regard to the thousands of other relations of life. Nevertheless this means that in this observer's role we can take in all

[3] As Marx writes in connection with the illusion of interest: "[A certain] portion of the profit ... appears to him [to the individual capitalist] to be the product [surplus value — M. A.] of a capital as long as it is operative; and this it is, as far as he is concerned, because he represents capital only as functioning capital. He is its personification." [K. Marx: *Capital, Op. cit.,* Vol. III, p. 373.] Similarly, a change of value of constant capital manifests itself in a different form, that is, in another conformation, for the individual capitalist and for the total movement. In the *Theories of Surplus Value* he writes that for the individual capitalist it does make a difference whether the increased productivity of labour takes place within his own branch of industry or within those which provide constant capital for his branch. But for the capitalist class, for capital itself, it is all the same.

external and internal relations and so can form a more objective and more correct judgment of them, whereas when we are committed to action we are dominated by the particular situation and so prevented from surveying things fully. Man is held captive by the situation which he has created himself. (The problem, however, has been alive in the history of philosophy ever since Fichte; he was the first to take cognizance of the autonomy of action as well as of the circumstance that genuine relations can be grasped only after the event and conceptually. It is a different matter that he conceived this in the category of 'intellectual intuition' tending to irrationalism.)

Immediacy creates objective spheres, the particular and common world of each individual person. This world comes into relative separation from the totality which it builds and by whose influences it is controlled, i.e. from the total movement of society. This is the dialectical relationship described by Marx in his theory of immediacy and illusion. In capitalism the individual's sphere of life becomes relatively detached from the practice of total society; in his particular quality the individual person begins to live a life of his own, establishing superficial contacts with other, likewise particular individuals. And the total movement, the class-like development not only in people's minds but ontologically as well is situated somewhere else. The sociality of man and his 'species-being' take effect only by the roundabout means of individuality or privatization. The coherence of these two poles is realized in the mode of existence of separation.

At the same time Marx always emphasizes that the point under discussion is not the contradistinction between the abstract individual and abstract sociality. Only definite historical conditions can generate the contradistinction, the superficial difference, between these two spheres. It is in capitalism that the particular individual's sphere of life most clearly assumes this configuration which turns everything around. For Marx, then, immediacy is a historical category, in contrast to Hegel and especially to the neo-Hegelians who saw in this pair of concepts (individual and society) a timeless phenomenon of consciousness or, like Max Stirner, a suprahistorical form of behaviour. It is in the worldwide phenomenon of privatization since the inception of Christianity in which Stirner sees the development of the individual sphere of life and to which he attributes the illusoriness (apparency) of the world. "Look out near or far, a ghostly world surrounds you everywhere; you are always having 'apparitions' or visions. Everything that appears to you is only the phantasm of an indwelling spirit, is a ghostly 'apparition'; the world is to you only a 'world of appearances', behind which the spirit walks." In the days of old there were gods behind things but now there are spooky, illusory beings: "But to you the whole world is spiritualised, and has become an enigmatical ghost ... a 'semblance' ... Since the spirit appeared in the world, since 'the Word became flesh', since then the world has been spiritualised, enchanted, a spook." [M. Stirner: *The Ego and His Own,* transl. S. T. Byington, New York 1918, pp. 36—37.] Marx noticed that immediacy has a history of development of its own just as the self-creating process of human praxis has. In the course of history various types succeed one another, and do so according to the stages at which the process of the socialization of mankind and the individualization running parallel to it have arrived. Immediacy for these two tendencies is a point of intersection of varying quality. Marx saw in the development of mankind a trend in the course of which man's socialization and universalization are inversely proportional to each other. Man becomes ever more a social being, objectively he is more and more imbued with social contacts, and simultaneously he also gains an increasingly richer individual life. But this individuality, instead of tending towards some inward enrichment, carries him in the direction of alienation and of the emergence of privatized individual. Immediacy is one of the objectifications of this inversely proportional development. It is a space for action and the plane of a picturing of reality on a distorted projection.

For this reason, only in the event of some definite social conditions will immediacy become a medium which turns everything around. In feudalism immediate individual activity and labour are not yet opposed to the total social activity. On the contrary, they are of immediate social value and reality. (In *Capital* Marx states that, since the basis of social conditions is formed by the relations of personal dependence, the products and labour do not have to take on various fantastic shapes which are different from what they really are. Here the nature, form and particularity of labour may immediately take a social form, unlike in the field of commodity production where they may fill this role by their generality.) Hence in this case immediacy does not as yet hide the connection between private activity and the total social movement, though the condition of this obvious connection is the restrictedness of the individual sphere of life. The individual person relates to the whole only as a member of his own corporation (through his priestly, military, knightly, nobiliary etc. status) and so the range of his possibilities for contacts are rather limited. His chances of establishing individual contacts with members of the other orders (of society) are still meagre and in certain directions are wholly ruled out. Consider, for example, Shakespeare who, while setting his characters of the nobility in a dramatic foreground which reaches out in many directions and with a high degree of referentiality, nonetheless presents his plebeian protagonists only in a single albeit rich perspective. Their completeness cannot even compete with the perfection of his important aristocratic characters. Notwithstanding these restrictions, it is true that private life in the modern sense appeared only around the 18th century. (It is no mere chance that Fielding is the first author to call himself a 'historian of private lives'.) Especially in the patriarchal period of feudalism, there hardly existed any sphere of immediacy in today's sense, that is to say, a contact zone separated from the interests of society as a whole. Even property assumes a form pertaining to the person in his capacity as such and not personal as in later times. The landed property of the nobility is inalienable. Should the nobleman sell his land however, he can retain his privileges which are provided solely for him, such as facilities for hunting or shooting. The new owner cannot possess these rights.

And it was this very same form of aristocratic ownership which bound the nobleman through thousands of personal transposals to the affairs of the entire society which he personally may also have helped to establish. This relation manifests itself even in the lower classes. It is aptly illustrated by Romain Rolland in the story of Colas Breugnon, or just take the legend of Joan of Arc where private action — with a certain tragic intensification in her case — conforms with an almost natural logic to the greater interests of the entire society or the emerging requirements of a nation. But even the evil schemes of Gilles de Rais, Bluebeard, these typically private crimes assumed a social character. They were followed by lengthy judicial proceedings because that was what was demanded by the outcry of 'public opinion'. This 'public opinion' expressed itself in similar forms, even though they were forms of false consciousness, in the hysteria of various religious movements, in traditional market-place games and in village communities. Thus, for example, as opposed to the cautious attitude of the official Church which even took countermeasures, religious mass movements mostly assumed exalted spiritual forms and turned into popular orgiastic cults which proclaimed the cares and passions of the social community as well as those of the religious one, and which also broke through the restrictions of ecclesiastical injunctions. The Church knew only too well that the appearance of eroticism would lead to the autonomy of emotional life and, through it, to the independence — liberation — of the commoners' way of life. (Max Weber clearly recognizes in these orgiastic endeavours the events leading up to the revolutionary struggles of the peasant movements.) The process of raising the erotic to the level of consciousness marks the beginning of a new phase in the peasantry's achievement of self-esteem. It is a germ of the transition from the simple state of nature to that of the fighting spirit, a transition which was at the beginning embodied by religious ecstasies, hysteria and orgiastic elements. It stands to reason that we are not talking about a 'public opinion' in today's sense, the diffusion and general acceptance of a dominant tone which has originated largely after the event, but about the common form of expression of a demand which is at the same time immediately communal as well as individual and personal. In the choral music of the 16th and 17th centuries we can still recognize and enjoy the reverberations of this immediate communality, while in the art of Bach and Händel we see the tragic reminiscences of it. The centuries-old store of melodies of the Protestant hymnody drew its sustenance from this immediately interwoven individual and communal existence and could so become the basis for the tremendous upswing of art music. To sum up: here private life could still be lived only in its personal and passionate richness, but this personal fulfillment could at the same time also be the medium of the historical interests of society as a whole. The possibility of an immediate unity between the two spheres is still open.

Marx, however, does not simplify the process of development. What he wishes to prove is not that these two spheres have assumed increasingly divergent dimensions, a process which then culminates in capitalism. That is why he points out that earlier in a slave society these two spheres would make contact only at the price of conflicts and that here, too, the immediacy of individual life — with reversed values — created illusions. As contrasted with capitalism where even overtime looks like paid work, here even the work necessary for the reproduction of manpower also looks like overtime, and these economic frames even narrow down the circle of possibilities for individuality. Then again, the exchange of goods and a money economy relax these forms, and it is the joint effect of these two tendencies which actually leads to the rise and to the golden age of *polis* democracy: to the relative independence of communal man, to the emergence of his individuality and to the priority of *citoyen*-existence. But the development of the late Greek drama — and of philosophy — aptly shows how precarious the balance between these two tendencies was and to what extent it was only through conflicts that they could assert themselves. The price of individualization was the gradual disappearance of the *citoyen* attitude, while the acceptance of the interests of the *polis* resulted in the tragic surrender of the emergent individual autonomy. Ágnes Heller presents an excellent analysis of how this irreconcilable contradiction is reflected as early as Plato's philosophy and how it becomes the basic material for the evolution of Greek tragedy. The appearance of private property, the progressive development in the exchange of goods made these communities a unified *polis,* while at the same time this tendency also prevented the survival of *polis* democracy.

Hence in the course of historical development immediacy, as a sphere of life, returns in a variety of metamorphoses. It assumes a different function in each social formation.[4] Its distorting influence will emerge only in the case of the interference of certain structural conditions. Its purest form is shown in the commodity production of capitalism where the order of the particular individuals' contact with one another becomes independent: the exchange of goods, bartering relations, the market, as well as circulation are all forms which are secondary in relation to the total movement but which for the daily life of people whose immediate motives and behaviour they determine are of prime importance. It is a medium in which the systematic arrangement of their reciprocal contact, demeanour, language, fashion etc. will take shape. Hence circulation is a model in which immediacy as an objective sphere can be studied as a commodity created by individuals and yet a medium grown independent of them, a sphere influenced alike by the total movement and by individual activity.

For individuals this medium makes each fact appear 'differently' and objectively presents it in a different form, thus becoming the midwife for the birth of illusions. That is why Marx lays particular stress on the fact that economic configurations which manifest themselves for and are used by the individual capitalists are not only illusory forms but also factually 'differently' existent. In *Capital* he writes that in the movement of social capital, i.e., in the movement of the totality of individual capitals,

[4] Moreover, Marx thinks that the false type of immediacy which is prevalent in bourgeois society will disappear again in a communist society, that it does not necessarily have to follow, as an abstract destiny, from the incomprehensibility of the division of labour or of technical development. He elaborates this in the *Grundrisse*, writing that by eliminating the immediate characteristic, i. e., the solely individual nature of living labour, or in other words, by ensuring that labour, as purely internal or purely external becomes universal, i. e., social, this form of alienation will break away from the objectivistic element of production.

things are represented differently from the way in which they appear for the individual capitals, as they can be seen from the point of view of the individual capitals.

Particularity, the immediacy of individual life becomes independent as an objective sphere. But this independence is by the same token a condition for the functioning of the entirety of social processes, of the invisibly working totality which is built upon it and which receives some of its impulses from it and also reacts on it as a determinant. Marx sums up this dialectic in the *Grundrisse* as follows: Simple circulation is a form of presentation in the sphere of the process of the total production of the bourgeoisie, which is the result of a deeper process working behind it and resulting just as much from it as generating it, i.e., the result of industrial capital. This dialectical relationship between independence and dependence characterizes the ontological plane of immediacy: it is independent and dependent, moving and moved at the same time.

However, what ontologically forms the unity of the dissimilarity in workaday practice and consciousness is only registered as a difference. Individual people can see only those forms of presentation which manifest themselves for their individual conditions of existence, and they cannot take in the contradictory connection of these forms of presentation with the complete whole because these two spheres are invisibly separated from and connected with each other by the dimension of time. The present time of immediacy shows an incessantly different aspect of the relationships, as the anterior and the posterior of the processes and so their totality in time. It is from here that there arises the other possibility of illusion. The reason why we tend to identify the forms of presentation with the — virtual — essence is because genuine relationships appear only prior to or after the event, while at the moment of action they are objectively inaccessible. We could see this even in our bus example: the prior story capable of 'explaining' the case was objectively excluded from the present time of the incident. The same applies to the relationship between immediacy and the movement of total society, with of course the added essential difference that, though the real elements explaining the facts are objectively excluded from the present time of practice, ideologically they are still supplied. Every society offers certain pseudo-explanations, false antecedents or motives which seem to re-establish the quasi-story and thus the quasi-totality of the facts. With this, however, the illusion of factuality becomes even deeper. This ideological complementation is, however, the subject of another chapter in this study.

Apart from this, the social division of labour has created a structure in which the individuals' deeds have a certain form and value at the moment of 'doing' and acquire a different form and value again when they enter the circulatory system of society. The respective levels of immediacy (the particular activity of the individual) and social praxis make contact with each other only after a certain time-lag, changing factually, too, the forms and result-related characters of things and deeds. That the deeds will have an outcome different from what we have planned on the basis of probability means also that the results achieved will also be changed in the course of social reassessment. (The achievement of private work takes on one form of value in the present time of the production and a different one in the social equalization of marketing, i.e., after the event; the specific product is measured according to the demand made and the amount of work invested by the total society, but this measure has no place *in actu,* in the present time of the activity, and it will contribute only subsequently to the form-determination of the activity. In the artist's studio the sculpture has a certain characteristic which is different from the one encountered subsequently by visitors to the exhibition. But the form-creating effect of the time-lag is shown even in our bus example: it is the earlier events which mould the immediate form of the relation arising between the driver and myself at some later time.) Due to the wealth of movements of the transposals in modern society, the unity of these two spheres can be realized only extended in time. The fact of 'subsequency' will transform the things I have created. They will return to me in a different shape and with a different value, and I shall be unable to decide whether their real form is what I had originally given to them or the one that they assumed subsequently.

This time-lag is both a condition and a consequence of the independence of individual practice, of the separation of immediacy. This phase difference conceals the connection between the two spheres: what occurs earlier emerges at a different time from what happens later. It is the difference in time-phases which separates — and with an objective force at that — the total movement from the sphere of immediacy.

The time-difference between the two spheres and their subsequently formed unity are for Marx an explanation for yet another theoretical problem: it is due to these factors that the objectification which characterizes all human work is distorted and assumes an alienated, fetish-like configuration. It is the objectification, the objective process of work, which sets the individual in the organism of the social mechanism, but it also conceals this socialization. Since this sociality emerges only subsequently and has no influence in the present time of the action and cannot be seen, it impels immediate action in precisely the opposite direction, towards privatization.[5] With this 'truer' degree of being, originating subsequently, there comes into being not only a fetishized objectivity but also a possibility for the creation of illusions. 'Subsequence' determines virtually, retroactively, in the consciousness of people. It reverses the one-way dimension of time and explains the changing result by the earlier starting-point whose form is different. (It happens on countless occasions in the fashion industry that with a relatively small investment and a slight utilization of value considerable success can be attained through the media of prestige, fashion-craze and manipulation, and that the illusion can be created that things which have subsequently become successful had origins of full value.)

What this means for the individual person is that the validity and degree of reality of present action are different from what will develop in a subsequent phase of the process realizing itself socially. Immediate praxis is in limbo between two degrees of reality: in the world of present 'making' where everything is tangible but where things will sooner or later turn out to be not real by virtue of the fact that in the meantime they will have been reassessed; and in the subsequently arising reality which is more spiritual in an axiological sense because it appears mostly as a quality of value, i.e., the sphere of results. Naturally some time later this subsequent result on the way to realization will also become a thing of the present, but for the man of immediate practice the situation will thereby become more difficult as he will not only have to choose between the 'two presents' but also have to be able to decide which of the two is the more genuine role. And in this conflict he is bound to choose the more spiritual result, the one which realizes itself subsequently. In this respect this decision still follows the ontological structure of the social movement, but conscious reflection by taking this decision also opens the way for illusions. For conscious reflection it seems that this socially transformed final result, the unintended configuration of another kind, is not only more real but that at the same time it appears to determine the motives of the individual also retroactively. The 'subsequent' results, those which

[5] In the *Grundrisse* Marx describes this relationship as follows: The emphasis has fallen not upon the objectifying being but upon the alienated externality, not upon the worker but upon the personified conditions of production, upon capital and its power. In this respect, the relationship between objectification and alienation is not merely an illusion but also an actual fact: this invertedness, this inverse relation, is real and thus not merely an imaginary one which appears only for the workers' and capitalists' mental conceptions.

are being shaped in the transition of the total social movement, are ontologically indeed more real than the results of individual and immediate activity. But this different result does not retroactively determine the conditions of its origination. (The fact that of the many different rival groups it was the Beatles that the manipulation of pop music chose, and, hence, the social accomplishment of the group's production has far exceeded their original talent and hopes, does not mean that by reason of their subsequent success their music has become better either socially or in value terms.) But in the sphere of the practical reign of subsequence that is what it looks like. This altered result will alter both the conditions of origination as well as the content of the original product, endowing them with a different form of objectification. The social determination of value, i.e. the form of presentation or the existence of validity, appears to be the genuine reason for the existence and as the substantial content of these very same things.[6]

Since the sphere of immediacy is perceptible at the point of departure manifesting itself for the individual or merely being 'made' in its altered quality as a result, this sphere is timeless, i.e. without a history. We can see only the naked facts but we do not encounter the history of their origination which might actually explain these facts to us. And they are determined by that which lies behind them (their historical development as well as the wider contexts which are also excluded in the 'single-channel' order of reflection of immediacy), transforming their manifestation for us into a different shape. Here there is light thrown for the first time on the deceptive nature of factuality. In immediacy there is but a single factuality, i.e., the datum which can be verified unequivocally: for example the date of a writer's birth and death. It will not do us a lot of good however, since data are mute in themselves. The greater proportion of facts such as we encounter in our daily life are in reality the petrified and objectified configurations of various processes, but these configurations also conceal their origins, the formative processes behind them, i.e., the possibility for an explanation that might bring to light their 'being so'. For example, it is a fact today that Modigliani is one of the most highly

[6] The problem of the new imaginary reality emerging in capitalism was raised first by Lukács in his *History and Class Consciousness* [*Op. cit.*, p. 257 ff.]. Though Lukács always emphasized that in this youthful work (which has an intellectual source of energy to this very day) the categories of reality, objectification and reification are shaded into one another, nevertheless imaginary reality gives a description of the objectification of time and of man's succumbing to this objectification and of his thinking in illusions, a description which holds true even at present.

respected painters. But the very same paintings which are auctioned at incredible prices today were of no value at all in the view of the general public at the time of their origin. Modigliani's talent and the quality of his paintings, i.e., the immanent factuality of his character, has not changed an iota — still, it is taken today as a different fact from what it was a few decades ago. In this case the content of facts has been changed indeed as a result of the joint influence of the evaluation of society and of manipulation. (In the Great Patriotic War of the dictatorship of the proletariat Jenő Landler wanted to introduce the use of the national colours in the Hungarian Red Army, but his enemies regarded this as nationalism because, factually, the opposing camp also used the same flag: the difference in content was not discernible.)[7]

We must emphasize, however, that the ambiguity of facts results only from the way in which they manifest themselves and not from their ontological content. They are the points of intersections of genuine processes which are illusory only in their immediacy since they block the prior history of their origin from consciousness. Such points of intersections that become facts are part of the nature of all processes. In both individual and social life there are events which turn a 'new page' in history. Once the alternatives of these events have been brought to an issue they will determine the course of development for some time. But at the point of intersection something happens which will end the earlier process and start a new one. Some major outside events (a war, a crisis, or the death of a member of the family) enters into people's life as such an objective 'wall'. These facts, however, mean one thing from the point of view of the summing up and closure of their own processuality and something else in the form of manifestation in which they enter into the life of people who live in other processes. The outbreak of a war is the

[7] This historicalness of facts was stressed already by the well-known Introduction to the *Phenomenology of Spirit*. In modern philosophy there has appeared the opinion which, by emphasizing this historicity, does indeed criticize phenomenology and neopositivism's category of 'factuality', but at the same time discards the facts. To quote Herbert Marcuse's words, for example: "Facts are what they are as occurrences in this struggle. [That is, in the struggle between man and nature. — M.A.] Their factuality is historical, even where it is still that of brute, unconquered nature. The philosophical concepts retain and explicate the pre-scientific mediations (the work of everyday practice, of economic organization, of political action) which have made the object-world that which it actually is — a world in which all facts are events, occurrences in a historical continuum." [*One-Dimensional Man*, New York—London 1964[1], 1968[3], p. 185.] In this case historicity manifests itself with an edge of attack with respect to neopositivism, but it gets stuck within the circle of everydayness and thereby limits itself.

'solution' of a lengthy political crisis. The outbreak is an objective fact in the same way as the process is in which it originates. But in the daily life of people it manifests itself not in this dual mode of existence but on the level of immediacy. It is only a fact whose causes, determinants and thus its content cannot be seen. Anybody can provide his own particular explanation for the factuality of this event and can pin his hopes on its particular outcome. Factuality is thus the way in which various processes manifest themselves for us. They form a point of intersection of reflection in a petrified configuration and are therefore full of possibilities for illusions.[8]

Add to this one more element of importance. Every fact is transcendental. They are determined and created not only by the logic of their prior history but also by the complex of circumstances in which this developmental logic could take shape. The fact condenses this elaborate complex into a single element and by doing so not only objectifies the movement of the ramifying relationships but also causes it to vanish. It often happens, for example, that a collision between two cars is caused by a third one which is not even there at the moment of the accident but which after confusing the first two overtakes them and disappears. The solid fact, the *factum brutum*, allows only the two colliding cars to be seen, whereas the mediation which has brought about the fact is invisible precisely by reason of the objectivity of the naked fact. The effect of the third car does not show on the two wrecked ones. Here, too, the 'fact' is ambiguous, and with this ambiguity it becomes the vehicle of the possibilities for illusions.

Facts are petrified processes.

By means of its industrial rationality modern capitalism has set up an order of manipulation that is capable of grasping and using these

[8] The fetishized relation to the facts is created primarily by the structure of bourgeois society, by the domination of immediacy. But its form of consciousness, the absence of a memory is already the product of the bourgeois outlook. Whereas the second wave of the Enlightenment and then early bourgeois thought strove to establish a historical outlook, the modern attitude is to try to get rid of it. Adorno, for example, formulates this switch wittily, though he absolutizes it with respect to the entire bourgeois rationalism. Thus he writes that the terrible vision of humanity devoid of a memory is not only a symptom of decline but it is in close connection with the progressive nature of bourgeois principles. Such economists as Sombart and Weber set down traditionalism to the social formations of feudalism, and ascribed rationalism to the bourgeois society. With this, however, they alleged no less than that progressive bourgeois society would eliminate memory, time, and even the act of recollection itself as some sort of irrational relics of times past. [Cf. Adorno: *Was bedeutet Ausarbeitung der Vergangenheit?,* Frankfurt/M. 1960, p. 14.]

petrified processes in their formal capacity, i.e., in their imaginary reality. Lukács considered that the superiority of bourgeois over feudal law consists precisely in its capacity for grasping this reality which is independent of these genuine relationships but has a role in its social functioning. [Cf. *History and Class Consciousness, Op. cit.,* p. 283.] He establishes the same thing with regard to the processes underlying the rise of bureaucracy and the rationalization of production. [*Ibid.,* p. 276.]

But individual practice also contributes to the creation of the fetishized form. It is an indispensable feature of our daily activity that we handle the facts without taking their origin into account. It is only in their objectified and fixed form that we can grasp the processes. (In official practice only those intellectual faculties can be used which are verified by documentary evidence, even if this certificate covers the actual knowledge only on the average of the cases and cannot even give an indication of the increase or decline of the abilities which have been acquired.) Institutionally it is only by means of factual data that the characteristics become apprehensible and usable. Qualities of character can be grasped only when they materialize in some 'tangible' gesture. The form of apprehension will indeed further the fetishization of facts: the manner in which we must grasp the processes is stereotyped until finally we lose sight of the notion that there are processes behind things. So the factual configuration not only hides the speciality of the individual processes but it also carries with it a number of illusory elements. But this structure is still in close connection with the functioning of everyday practice based on the 'probable percentage of hits': this is the only way in which we can work fast and objectively at that. But the same structure is utilized also by fraud, i.e., artificial pretences. Hypocrisy, forgery etc. all rely on the ambiguity of facts.

The latter examples are contingent ones. Ontologically, facts originate in a process of transformation in which the total movement is pictured on to the plane of the individual sphere of life. That is why Marx says — as we have already quoted — that the configurations of the total movement of capital appear factually, too, in different forms for the individual capitalists. These configurations become illusions when the reflection attributes this immediate form of manifestation not to its ontological mediatedness but to its conceptual and subjective explanation.

Now neopositivism and phenomenology insist on this essentially process-denying factuality or are at best under the illusion that on this basis they can reconstruct the heterogeneity of creative processes. In reality, they remain within the sphere of the immediacy of factuality notwithstanding their best intentions. As Kierkegaard put it, it is only by means of immediacy that they succeed at best in breaking through the sphere of immediacy. Adorno, in his critique of Husserl, pregnantly demonstrates the connection between phenomenological factuality and fetishizing. He says about Husserl that his infatuation with production leads him to the point that he regards the product as something fixed. Even the most extrinsic abstraction will unconsciously be guided by the general tendency of a society which cannot but present its existing reified forms as conclusive and categorial. With Husserl, it is in the innermost cells of epistemology where an attitude appears which fetishizes the already existent. [Cf. Th. W. Adorno: *Zur Metakritik der Erkenntnistheorie,* Frankfurt/M. 1956, pp. 151—152.] Adorno's slashing criticism is partly to the point: factuality is indeed the result of a dual process. It is the result of the immediatization and transformation of the total movement of bourgeois production and of the fetishized way in which consciousness sees things, while of this latter factor the phenomenologist notices nothing but the mere result. Husserl himself, however, comes near to admitting this narrowness of his approach. In his later work he himself exposes the inconsistency of the theory of factuality.

In the *Krisis* he aptly describes how the modern fetishization of facts — the factuality of neopositivism — has alienated man. He writes that the exclusiveness which in the second half of the 19th century determined the modern man's entire world picture and by whose 'pureness' he allowed himself to be deluded also meant that he apathetically let himself be diverted from the real questions that were of decisive importance for humanity. Purely factual sciences create purely factual people. [Cf. E. Husserl: *Die Krisis der europäischen Wissenschaften und die transzendentale Phänomenologie,* The Hague 1962, p. 4.] There is an ironic overtone in this fierce critique, namely that in spite of Husserl's good intentions it also applies to his later attempt at establishing a system. The subjectivistic reconciliation of factuality does not do away with the fundamental position of neopositivism. As a matter of fact, it is with this brilliant attempt that the ways of modern philosophy part once and for all. One method is to relentlessly criticize immediacy, but this method uses sociologically motivated arguments to sceptically deny the transcendency of immediacy, or holds out the prospect of only relative successes. Adorno, in his *Negative Dialektik* [Frankfurt/M. 1968], finally adopts this attitude. Though objectivity, the object-being, is not identical with the 'reality' of immediacy, nevertheless the progress towards real 'things', i.e., relationships, means only the re-creation, and indeed the negative re-creation of immediacy. What becomes for man a usable immediacy is not a sphere of essence as Hegel and Marx suspected. Man is surrounded by 'wrong immediacy', by the endless self-repetition, with its immediacies enclosed in ever new masks. [*Ibid.,* p. 181 ff.] Sartre transcends immediacy with all the destructive vanity of a Herostratos: man, by accepting liberty, by the anarchistic or heroic development of the ego, can break through the wall of immediacy or at least pass beyond this sphere on an ethical plane. Lukács, however, elaborates a theory in which he can reconstruct the Marxist concept of immediacy in response to the questions raised by modern development. Lukács is aware of the suffocating role of immediacy just as Adorno or Sartre are, but he also knows that transcendence is primarily a matter of social praxis and that in the last analysis it is a question of political struggle, social evolution and total social movement. So far as any of this concerns the sphere of action of the individual person there are several opportunities for transcending immediacy. One of them is obvious, namely scientific analysis or the elucidation of the superficial connections of reality. But the ordinary person rarely lives in the world of scientific achievements. For him even the most revealing scientific discovery may become an immediacy again, and a wrong one at that. Accordingly, there is a different opportunity available for everyday life, namely art. But although art re-creates the virtual picture of immediacy through the world represented, this is already a compatible immediacy, a world in which the genuine relationships are also perceptible and where the sphere of data is only a 'guideline', a pseudo-reconstruction of the original, fetishized and illusory immediacy. In reality it is the exposure of this immediacy. So art is just as legitimate a breach of immediacy as scientific thinking is, and it would be a scientific illusion to believe that one can be delivered from the illusions of modern life, from the elemental flood of immediacy, simply by scientific proficiency. Any significant work of art demolishes this sphere and helps man to truly discover himself and to turn the illusory sphere towards reality.

2.3. OBJECT-MAKING, EVERYDAY OBJECTIVITY, AVERAGENESS

Immediate practice works with two different forms of objectivity without being able to differentiate between these two configurations both of which are of objective quality. One is the original and natural characteristic of things, whereas the other one is the teleological or social function based on that characteristic and of equally objective validity. Even work and play processes are characterized by this dualism. A case in point is the use of the two-armed lever. An iron bar or a pole becomes a lever only in its specific mechanical application, i.e., when it is used in accordance with the functional rules of its being 'in the nature of a lever'. Apart from this functional relationship the same article is only a piece of iron or wood. In order to lift a heavy load by a slight application of force I must take it in my hand as a lever instead of as a pole. I must properly select its fulcrum and the siting of its load-lifting end etc. At the same time I must also take into account its original and natural characteristics, for example that, among other things, I cannot lift as heavy a load with the pole as with an iron bar, because the wood would break. In fact, I handle simultaneously both its natural characteristic and its teleologically applied function. On the other hand, once one has successfully applied it 'as such', one is inclined with natural naiveté to identify the iron bar with the role of the lever. Even in its 'being in itself' the object is for consciousness identical with what I have used it as, notwithstanding that in the course of practice I clearly felt and saw the difference between the two kinds of objectivities. (I knew, for example, that it would work as a lever but would break as a pole.) In the course of our daily practice these two kinds of objectivities are increasingly interlaced and for consciousness become identical.

In the first step this identification is furthered by teleological directedness. For, it is obvious that in the end-oriented activity it is not the object itself that we grasp but the objective possibility (and its ontologically inherent potentials) and 'make it our object'. It is not the thing in its entirety that we grasp but only that which will 'become' for us from this thing. Practical interlacing is thus preceded by the view displayed by teleological consciousness with respect to the object. If we perceive the thing at all we do so from a certain selective point of view, according to its future utilizability (this is likewise a form of object-making), and we shall work with it on the basis of this potential characteristic, gradually realizing what we need and what was ontologically indeed inherent in it as one of the many characteristics of the object. Gehlen has a characterization of this anthropological feature. He writes, for example, that the purpose of thought in the rational knowledge of objects is not to approach an already existent reality; on the contrary, it wants to stress those characteristics which are not the inert capacities but the dynamic relevancies of things and which it considers with respect to the possibilities into which they might be changed in the course of a supposed operation. [Cf. A. Gehlen: *Der Mensch, Op. cit.,* p. 319.] The practice of object-making handles relatively separately the particular and the instrumental nature of things, whereas practical consciousness, i.e. teleological reflection, does not make this differentiation since it views things with respect to the realizable possibilities.

The same identity and difference can be noticed in children's games. When children are playing with sand biscuits they work with two different qualities simultaneously: with the wet, thoroughly kneaded sand and also with the 'biscuit' which is supposed to be eaten by the guests according to the rules of the game. But here the two planes are still distinctly separated. Even in the heat of the game the natural basic material and the superimposed function which has been 'made into an object' are already separated from each other, and the thought of actually eating the sand would never enter the mind of any of the children, although the very object of the game is the biscuit and not the sand. This dual objectivity in the game is of course the childlike reproduction of the teleological working process. What becomes entwined in the world of adults remains still separated in the playground.

But, as we have said, not even the working process can fully identify the different planes of the objective and the functional existence. Not only the success of work but teleological consciousness is also subject to the 'interlacing' of the two different objectivities. This identification arises mostly in social practice and consciousness, where the natural property of the object and its functional existence are built on one another in a much more fluid state. Remember Marx's classical example, namely that the natural properties of gold and its function as money are founded on each other. It is the objective quality of gold (it is inoxidizable, rare, occupies little space, represents great value and is easy to divide) that makes it suitable for fulfilling its function in society, its role as a means of exchange and as a measure of value function. But its second capacity, as money, though of social origin and merely based upon its natural attributes, has just as objective an existence as its original properties, the only difference being that this objectivity has a different dimension. Marx sets store precisely by the paradoxical process in which gold, as a means of exchange, figures only as a symbol of representation and so its own original and natural properties not only become a matter of indifference but may even disappear altogether (for example, in the case of the banknote). But material existence concealed behind functional existence is still present: only through making an object of its material nature can this function of gold be grasped and used. In other words, it must be collected, exchanged and circulated as a metal. Consequently, what seems unrelated to its material quality in its functional existence can be brought into action only through its natural objectivity in practical use. These two different kinds of objectivities are merged into each other by the performance of practice, i.e. object-making, by use. 'The object is identical with what I use it as.'

Marx also refers to this practical requirement, to the substantializing effect of object-making. In the *Grundrisse* he writes that the exchange value may naturally exist only symbolically, though it must be endowed with an objective existence if this symbol is to be used as an object and not as a mere imaginary form. The exchange value is not only an ideal notion but it is also an objectively pictured datum. We can hold the measure in our hand, for indeed the exchange value does measure but it will exchange only when this measure passes from one hand into the other. Hence it follows from this statement of Marx's that reification is the unification of an ('ideal') function, which has a genuine though abstract existence, with a material vehicle in the course of 'apprehension', i.e. object-making.

Practical identification, as we have already said, can never be complete. It is carried out by everyday consciousness which forms a definitive unity out of the possibility of practical identification. To be more exact: consciousness proceeds simply from the functional, i.e. socially operating existence and conceives of this as the characteristic of the original and natural objectivity. (From ancient times to the period of the Enlightenment there had been a struggle going on in which writers and philosophers fought against the natural properties of gold without realizing that the material features were merely the vehicles of a secondary objectification which was based on them, the vehicles of the equivalent form of the socially developed exchange of goods.) So the act of object-making only prepares the way for this identification, offering reflection a possibility in which both the natural and the social determinations are handled as a single unit. Consciousness — since no other form of approach is available — accepts this 'phenomenon' and registers it as an illusion. (In the history of civilization, for example, a long time passes before the linguistically grasped and thus formally objectified things are separated from the genuine objects. The joy of linguistic apprehensibility and expressibility is for a long time objectivized as if we were actually holding things in our hands: the magically uttered words substitute for the objects themselves.) In the process of object-making there are three different kinds of objectivities entwined within one another and they become virtually a uniform objectivity. While one layer is the natural structure of things (the pole in the case of the lever and the sand in that of the game), the second one is its teleological and instrumental application. I handle the object in accordance with the prescriptions of a certain function, without making use of all of its characteristics. And finally, every socially developed instrument is also the vehicle of certain definite social relations, for indeed it can function only if at the same time it mediates the class relation between people. To demonstrate this by an example of Marx's: a particular machine in a factory is the vehicle of a social relation, of exploitation, of the production of surplus value, of class antagonism. Its technical and mechanical materiality is at the same time a vehicle of a social relation which is, however, not resolved in it but is merely built upon it because the machine can function in its social objectivity only if it is used under definite conditions, i.e., if it is made into an object in the capitalist way. It is only scrap iron if, on the occasion of a shipwreck, it lands on a desert island, but it becomes that even if, on account of its growing obsolete or due to its amortization, it is dismantled and sold to some private individual who will use it perhaps for some other purposes. (For example, as a component of a 'mobile', a kind of modernist sculpture.) The reason why a social relation cannot in itself be grasped is because it has no tangible objectivity; nor is it visible, though the physical and technical structure of the machine is also determined by this relation. (The level of technological development has been shaped by competition, technological revolution, the state of the labour market etc.) In social life an ontologically superior level of objectivity takes effect but we can sense its constraint only after the event, in the successes or failures (this constraint has no immediately tangible form), or in the objective vehicles to which it is linked and which convey this objective relation (machines, institutions, natural attributes such as instruments and the like). This higher-order, social objectivity, i.e. the compelling force of relations, exists just as much as the natural environment or the functional possibility of the instruments to take effect does, and we obey it when we are using it just as we must obey those social relations which are conveyed and mediated by the instruments. In object-making these three planes of objectivity take effect simultaneously and the difference between them will also disappear. Either the role as a means hides the specific objectivity of the naturally objective as well as that of the socially objective, or we can heed only the materially apprehensible objectivity (as in the case of money). We are unable to survey the structure of identity and difference of these three layers because it is through the use of the material and instrumental functions that in actual practice we bring the mediated social objectifications into actions, without having to be aware of these factors. The self-justification of illusion is created in object-making.

Object-making is an elementary form of objectivity in everyday life. It exists in many different forms. It can be performed linguistically as well as in work, ethically as well as philosophically. (In the pity and compassion I feel for someone I also subsume the other person in my own ethical value.) Each performance of object-making grasps things in a different dimension, in a different aspect, and it is also conceivable that the very same elements may exist simultaneously and yet separately from one another in their object-making of various dimensions. (For example, I may like someone as a boss and yet find him insufferable as a person. I make him my object in different ways in the two relations.)

In reality this act is of course nothing but a final result. We use things 'in just this way' only as the result of a lengthy social praxis, as the realization of its preliminary process. Since, however, it is we who make them 'like that', so in consciousness the individual realization is stylized into the creation of the characteristic feature of objectivity. (Schelling's illusion of the 'creative intuition' is the idealized — and false — reflection of this practical performance. In reality it is not we who create the objects like that, but in the immediacy of individual praxis this performance seems nevertheless like the source of the 'being so'.) Object-making, therefore, characterizes man's immediate relation to things in an extremely wide range, from elementary working processes to the most subtle intellectual work.[9] Max Weber, for example, cites British jurisprudence to prove that the 'being so' of the legal principles is established by individual and contingent application, i.e., that any legal principle exists at all only in the form in which judicial practice 'makes it its object', in other words, applies it to its specific case. Weber writes that even Blackstone calls the British judge a legislative oracle and that, indeed, the part which the judge plays in his decisions as the indispensable and specific forms of the Common Law is comparable with such an oracle. What was previously uncertain (the existence of the legal principle) has now been changed (through the decision) into a long-lasting rule. [Cf. Weber: *Wirtschaft und Gesellschaft,* Tübingen 1956, Bd. II, p.

[9] Marx ascribes the consolidation of the illusion of the so-called 'triple formula', among other reasons, to this performance of object-making. In *Capital* he writes, for example, that it is quite natural that the true characters of production feel at home in these alienated and irrational forms such as capital and interest, land and rent, work and wages, because they were formed precisely by the illusion in which the characters of production move about and with which they are in daily contact.

407.] As there may be many different interpretations of a given principle, particularly in the British 'precedential law', and consequently there are various possibilities for object-making, so the decision will identify even the universal validity of the given legal principle with a particular functional form. Thus object-making endues an abstract intellectual configuration with an ontic form by a particular application of it.

The same structure can be seen in our active relations to institutions. Institutions offer and indeed impose on us certain behaviour patterns and stylize our teleological directedness insofar as they induce us to assume certain roles and modify our actions accordingly. It is the traffic rules which change us into pedestrians and incline us to treat others also as pedestrians. In the institution of marriage the husband's sphere of action is limited by the historically given possibilities which define the subordinate and coordinate relations of cohabitation, i.e., the being of the couple 'for each other'. In medieval customary morality the woman as chattel (for man's pleasure) was a social 'phenomenon', just as placing women on a pedestal is a feature of American culture. But from the individual person's viewpoint this objectively developing historical relation is only a possibility that is realized in the act of object-making. For a man a woman exists in whatever way he can 'make her an object'. An official, a sergeant or an interrogating torturer also sees his patient or victim as being whatever he treats him as. In the act of object-making the outward form, which for him is socially 're-formed', attaches itself to these people who have, for that matter, a good many other characteristics as well, and excludes these other features from his scope. Institutional forms of action operate with this secondary objectivity, i.e., with an alienated form of manifestation which is corroborated by the act of object-making: they are like that because they can be treated like that. This type of objectivity therefore also has an ethical projection, such as the sense of dignity. Last but not least, it is due to the reflex of the particular experience that man is usually judged according to the way in which he allows himself to be dealt with. Anyone who sets a small value on himself — however worthy he may be — seems to be of no value, while moral stamina commands respect even if it seems to be insufficient against sheer physical violence. These ethical reflexes are, however, merely forms of self-defence in a medium which develops socially and which the individual realizes. Ethically it is only by means of individual gestures that the more generally prevalent illusion to which this gives rise can be dispelled.

This act is, however, capable of treating as reality even fantastic and irrational forms of manifestation or configurations where fetishizing seems to have taken on, as it were, an unobjective form. C. Wright Mills raised the question of how 'superiority' appears to workaday American people and of how the 'manager' can be made the object of action. "Seen from below, the management is not a Who but a series of Theys and even Its. Management is something one reports to in some office, maybe in all offices, including that of the union: it is a printed instruction and a sign on a bulletin board; it is the voice coming through the loudspeakers; it is the name in the newspaper; it is the signature you can never make out, except it is printed underneath." [Mills: *White Collar,* New York 1956, p. 80.] Object-making is related here to elements which are alive only in their functional quality. The natural objectivity on which they are based has already disappeared. Management is not denoted by a human utterance, face or gesture but only by an illegible signature or a newspaper item. But this phantasmal existence comes to life in object-making because we are related to it. We accept it as such, since everyone treats it as such. So it is through social universality, through many individuals' accepting them and passing them on, that the fetishized forms of presentation attenuate into mere functions behind which the primary objectivity to which practice must be related, as it were, hardly even exists any more.

The most fetishized form of object-making is the bureaucratic relation. Its basis is an ontological structure, namely the interconnecting organism of the institutions of society. But its form of consciousness is the objectified outlook of people and of their personal and individual characteristics, provided that the mobility is missing. "To the bureaucrat, the world is a world of facts to be treated in accordance with firm rules... Research for bureaucratic ends serves to make authority more effective and more efficient by providing information of use to authoritative planners." [Mills: *The Sociological Imagination,* New York 1959, p. 117.] Individuals here are only numbers, nameless items or sheets of paper awaiting decisions because they have to treat them as that day after day. Even the inertia of practice forms this apathetic reifying attitude which can expect but little help from sociological research trying to improve these relations. (It is quite a different question where the socially necessary proportion of bureaucratization separates from 'Parkinson's disease', when bureaucracy overproduces itself, resulting in a state of inertia.) At the same time, however, this bureaucratic outlook and type of practice not only takes effect with respect to others but it also reacts upon the acting person himself. The diabolic thing in the tragedy of people like Ivan Ilyich is that, after having spent their lives in changing others into files, they also lose their own essence in the process. In the fetishized form of object-making man also loses his own true objectivity. If the only way in which the world can become an object for us is in its formal quality, we become even for ourselves chimeras which lack a core or possess merely an illusory essence. Marx writes in the *Economic and Philosophic Manuscripts* that only what is an essential object for man, i.e., his objective essence, is truly an object for him. In the bureaucratic relation one does not achieve the objectivization of one's own individual character. One does not objectify the essence of one's own self since one is unwilling to grasp the individual essence of those one is dealing with.

We encounter this vicious circle not only in the sphere of alienated authority, but the formulae of the microcosm also present this pattern. The spontaneous illusions of self-delusion and self-deception are also created by the (undetectable) bureaucratization of everyday conditions. O'Neill, in his play *The Iceman Cometh,* reveals that each of the *déclassé* characters living in an alcoholic stupor and in self-deluding illusions has lost himself in the process in which he had bound his companions and his surroundings to himself through some kind of a deceptive shallowness, whether as an anarchist politician or as a bickering husband. And their self-deluding illusions represent the morphine for their attitude in losing the world and, with it, themselves. That is why we regard Lukács's study *Néptribün vagy bürokrata* [Tribune or bureaucrat] as the fundamental situation-diagnosis of the age, a study in which he analyzes bureaucratism not only as an 'official' relation but also as one of the overall tendencies of modern culture, pervading private life, individual psychology, public thinking, artistic outlook and the intellectual spheres alike.

Thus the essence of object-making is that everyday practice and consciousness cause social and teleological determinateness and the original and natural objectivity to merge into each other. On the illusion of the 'triple formula' Marx says that the immediate fusion of the material relations of production and their historical and social determinateness have been completed *[Capital]*. We grasp the things in their natural quality but we can use them only in their socially formed determinateness. Object-making is the unity of these two performances (grasp and use), a unity which practical consciousness will identify in an illusory way.

With this, we have already touched upon the structural basis of object-making. Ontologically we deal mostly with relations, relationships and dynamic processes, but everyday practice is unable to grasp anything which is relational. It can handle only their objective precipitations and their substantialized forms. And moreover, the structure of modern immediacy will crystallize the relations among people in the points of intersections of things and objects, entrusting the latter with the mediation of the former. Thus in object-making we in fact objectify mobile relations which in their original nature are intangible. Hartmann has an apt description of naive, everyday consciousness. He writes, for example, that it grasps the existent primarily as things. Relations, movements and processes are less existent for it. It substantializes the things to which it ascribes a substantial character. It hardly ascribes any existence to the changes and the relations of their characteristics. It conceives of these relations as ontologically secondary and as the 'how' of substances. [Cf. N. Hartmann: *Zur Grundlegung der Ontologie.*] And this consciousness will ask with reason whether a brook or a forest is a 'thing' at all and whether, in the case of the brook, only the water is a thing, or the babble, the wave and the foam are also things. But the snag is that what Hartmann correctly establishes in connection with the substantialization of natural processes is partly an objectivization of the practical necessity of grasping. Mobile processes cannot be grasped. Substantialization is nothing else but the intuitive and practical form of object-making. All working processes would be impossible were we not able to objectify the chemical or physical processes as the characteristics of things, to shape them as such, and to translate them into some new teleological relationship.

On the other hand, this substantializing outlook occurs in a somewhat different way as regards social processes. Natural relationships can be grasped as ontologically independent characteristics, as objective appearances. Substantialization does no violence to objects. For one thing, certain processual configurations represent such a gradual change that for human existence they can be regarded as inert things, rather than processes. On the other hand, the objectifying and reifying performance of work also contributes to this anthropomorphic quality insofar as it stresses a particular real, objective (apprehensible) element of things and processes. The brook rates as a thing for the miller and it functions as a process through its tangibility: it drives the mill-wheel. Substantialization lends only a frame to the processes which are grasped as things. But the natural processes become substantialized in what is still a relatively translucent way: object-making lends only a teleologically utilizable form of motion to the process. And moreover, even if consciousness sees objects as primary existents, neither their relationality nor their motion will disappear in this frame but will 'dissolve' into something else and become usable. The situation is different in the case of social relations and their forms of presentation. Here, substantialization is no longer only a matter of tangibility. It is, of course, that too. The love-affair, as a process, as an emotional attachment which develops and 'floats' between two people is even in itself intangible because it is inconstant, mobile and turbulent, and it can be enjoyed and experienced only as the other person's thought, emotional expression or body, i.e. as his or her objectivization. But the social processes of motion are 'objectivized' even apart from the act of object-making and they change into real features or traits of character. The judge's dignity and his manners are consequent upon his official routine and position of authority, but to his acquaintences the same behaviour pattern seems to be a trait of his personal character. Football stars are the vehicles of a complex social relation: the public, the individual abilities displayed in the game, the *esprit de corps,* the fans' adoration form together the relation which will be substantialized into the real and the apparent personal characteristics of a particular star player. (The successes of the entire team enhance, as it were, the individual's trait of character, and what has been developed as a result of joint achievement or earlier successes is attributed now to his own person.) The relation to the stars cannot, in itself, present itself. This is even more the case with the 'appearances' of the economic structure. The total movement is pictured on to a single form of presentation and becomes the property of certain objects. In the act of object-making therefore there is an objective and socially occurring process of substantialization which coincides with the necessity for practice to grasp, and these two together constitute the form in which the processes disappear, leaving only the 'things' in our hands. The abstract possibility of fetishizing already appears here. It is one of the most important features of object-making, a feature without which everyday praxis would be paralyzed. Practical consciousness which orientates itself only to this substantialized objectivity will inevitably arrive at false relationships in the picture it forms for itself. The true relationship will also be included but in a short-circuited, unmediated form.

Everyday objectivity can, however, manifest itself in two different forms. One is the unalterable datum, simply commanding acceptance and conducive to the world of fetishization and alienation, whereas the other one is the world of things made into objects for us in the alternative choice. Let us have a closer view of the nearest structure of these two different kinds of possibilities.

We have already dealt with the phenomenal form of 'being for us'.

Since they manifest themselves for 'our situation' as 'phenomena', they are socially preformed configurations, but only within the bounds of possibility. Only in our hands, in the act of object-making will these possibilities turn into realities. The reason why we deem this important to emphasize is because the configuration of this manifestation represents an alternative possibility for praxis. The same data which manifest themselves for our situation as 'phenomena' may result in different decisions through the transposal of individual aims and conditions of character, i.e. we can make these 'phenomena' our object in a variety of ways. To be more exact, the 'being so' of things which have been made into objects depends on this alternative decision. It must be emphasized, however, that even in this case it is not we but our teleological directedness which creates the ontological quality of things and relations, and only then, in the course of our decision, will we identify ourselves with the objects in a particular mode of being. For example, the river exists when we intend to fish in it, or when we take a trip to its banks for the sole reason of admiring the view, and even when we want to cross it. But it owes the different structures of its given nature and of its objective form to the different structures of directedness. It will determine our actions in one way when we just want to fish along its shore and in a different way when we want to cross it, for example, in the case of war. Directedness lends even in itself a certain mode of 'being for us'. In this process objects will change from mere existents into tangible data 'for us'. But this datum, if we take social activities for our basis, is further differentiated again. One has alternative choices in dealing with it. It is true that the river manifests itself with a different objectivity for crossing it but, except in deadlocked situations, I can also choose perhaps a different possibility. I can either consolidate my line of defence here or I can execute some diversionary troop movements further up or down the river, and so forth. In social practice many of our activities give the illusion that the game can only be played with predefined moves. We get this feeling particularly in the case of routine, daily recurring working processes where the same thing seems to happen over and over again. However, it is up to us to choose, and even within the narrowest range of possibilities it depends on us what we will do at a certain point: whether we shall rebel or continue to discharge our duties, step out of the routine or do what is the correct thing to do. For indeed, even the routine itself is a succession of only outwardly similar actions, while at closer examination it turns out to consist of a series of minute decisions which are invisible even to us. (Our morning ride on the bus occurs day after day in the same way, but our adventures are not even twice the same. Every now and then we fail to board the crowded bus, whilst at some other time we insist on our rights somewhat more aggressively in shoving aside our elbowing fellow passengers, etc.) Objectivity is never implied in mere acceptance and still less in an *a priori* given reality. It is created in the alternative act of object-making. 'Being for us' is not simply a passive relation but it is an objectivity that we have also created because we have chosen it. The social tendencies or the proprieties which are binding for everybody come about by more and more people choosing this particular solution, and the individual person to whom these choices manifest themselves in their accumulated form as ready-made stereotypes can either follow or refuse them. The objective fixed points come into being in this alternative series of choices. Avant-garde music is made into an object — in an intellectual sense — in a different way in a medium where it also carries the values of 'forbidden fruit'. Here the element of choice has a clear-cut moulding power. It complements the aesthetic quality of music. It was the same with us, too. When this sociological 'plus' had disappeared and avant-garde music became a familiar phenomenon in the concert halls, the element of choice lost its importance and, with it, the quality of value of the music also lost that something extra which had been imparted to it by fashion or by the oppositionary attitude demonstrated in the choice. In both cases object-making played the part of a form-determinant, and it was only through this choice, through the creation of the new alternative and the social accumulation of individual choices, i.e. by becoming a style, that the music's aesthetic 'being for us' came through to us in this particular quality. The many individual choices have kept on accumulating till they assumed, as it were, an imperative form of attitude: the 'tyranny of style' is the consequence of this series of accumulating individual decisions.

It is interesting that in Husserl's late work, *Die Krisis der europäischen Wissenschaften und die transzendentale Phänomenologie,* a book to which we have already referred, he aptly observes that so far as we regard data as mere facts, i.e. not as alternative possibilities, there will remain hardly any difference between that kind of phenomenological outlook and the attitude of the bureaucrat. He writes that as men with a calling, we are unaffected by anything else, and we perceive only that which is the horizon of this world of ours and which is the reality and the possibility of this horizon, i.e. the world which is real and truthful in this state of goal-directedness. And he adds that the structure of this professional consciousness of purpose and interest is precisely the same as that of the phenomenologist's. [Husserl: *Die Krisis der europäischen Wissenschaften und die transzendentale Phänomenologie, Op. cit.,* p. 459 ff.] As soon as we abandon our alternative relation to objectivity we face a fetishized world. And — as Husserl divulges — phenomenology works precisely with this 'ready-made' fetishized reality.

Everyday objectivity is formed as the social average of alternative decisions. Its form-determination is given by this structure of choice, but after it has grown into an established tendency, it offers itself for the individual person as a spontaneous possibility which is worthy of imitation. We shall come back later to discuss the ontological structure of the average and the sham possibilities it carries. For now we merely want to point out that the process of fetishization is the obscuring of the alternative possibilities and the development of a monolithic obsessional psychosis. When objects, people and relations manifest themselves in their sole obligatory form, and we accept this, as it were, in a state of somnambulance, the possibility of deciding in a different way does not even occur to us. Brecht's *Mother Courage* does not recognize that there is an alternative to her way of life and that she is pushing her cart in a false circle, namely the same war that she has been living off has also robbed her of her sons; but this inconsistency is concealed by the illusion of immediate utility. The point that Brecht stresses here is the fetishization of consciousness. The components which make up the style of living do not reach consciousness. Dürrenmatt's scientists, in *The Physicists,* intend to hide the formula for the destruction of the world in a mental hospital and think that this is the only possible alternative. Their tragedy is that they have been deceived from the first moment. The manageress of the asylum has already expropriated their secret. Though the fetishization of consciousness and outlook here, too, helps to obscure the alternative possibilities, it is nevertheless a more objective process that totally conceals them. What also becomes evident from these examples is that there are various ways in which different versions of object-making (linguistic, scientific, ethical or labour-related) translate reality and transform outlooks. The gesture of linguistic apprehensibility and objectification may be fraught with problems (linguistic philosophy searches for the relevant inconsistencies in a subjectivistic attempt at a solution) just as the working process is always objective in itself, and what is fetishized is only the ideas about it, the teleological steering.

The ontological basis of this relation of alternative object-making is the circumstance that only on rare occasions does the objectivity of social conditions manifest itself in advance. Of course, the adult person generally knows in his routine actions what he can or cannot do. The bounds of conditions, i.e., their objective possibilities, surround his daily routines as some sort of protective walls of awareness. Our social rules, ethical standards, official duties and daily discipline rest on these inevitably known and observed frames. But beyond it, in the deeper layers which pertain to everyday life but which have a significance beyond themselves, one cannot act in this foreknown way, especially not in the relationship of the present and the future, in our teleological decisions. Someone who buys a plot of ground is not buying the piece of land which will lose its value in a few years or will be nationalized, but the small lot on which he dreams to build up his weekend cottage. He will make it his object as he has teleologically chosen it. And at this stage uncertainty will enter the picture, for indeed though the buyer of the lot also contributes to the socially active tendencies, nevertheless the other buyers and speculators, government measures and agricultural policy etc., will have a much stronger share in the subsequent fate of the plot he has acquired, and the objective conditions which control and determine its fate as to ownership will take effect only subsequently, in the ensuing result of object-making. The situation is the same with the simple decisions of everyday life. The buyer of a car does not realize that by this he may well become not only the owner of a vehicle, but also a slave to social conditions. At the moment of buying it he decided between alternatives. It was he himself who made the decision between buying a car or a flat. But his decision was only expedited by the fact that his performance of this objectification was not affected by those other, for him unfamiliar, objectively cogent and only subsequently active conditions which he purchased along with his car. (He has also made his object a particular mode of life, a tendency of demand, and moreover, the car might 'want' to go to some place, for example, where its passenger may not have the slightest inclination to go, while its owner falls into prestige-seeking habits which have not hitherto been characteristic of him, etc.) Such 'concomitant' conditions governing the life of a motorist are, however, not the inevitable results of 'making an object' of cars. It is here that the continuous self-manifesting alternativeness of object-making presents itself. There are always two options for the owner of the car: either he yields to prestige consumption and, with it, he alters his mode of life and distorts his individuality, or he treats it as a mere conveyance and frees himself for the additional enrichment of his individuality. Hence being a motorist represents a specific condition of life only in its tendency. It rests with the individual's ability to choose between alternatives and with the staying power of his individuality to decide which course he will take. And this alternative is present — at least as an extremely limited abstract possibility — even in the case of the most fetishized relations.

It is at the same time undeniable that in object-making, as an immediate practical relation, there is a certain spontaneous temptation, namely people are under the command of things. Social relations are built into objects and by presenting a certain probability they drive one's actions and thwart one's claims to make an alternative decision. Though we apply this probability as a subjective category, it is not that. We would be unable to take any step at all if we could not, with a fair degree of stochastic probability (but not as a dead-certain inference), judge the braking distance or the speed of the oncoming car or the likely consequences of a gesture, even when acting on these data of probability becomes a habit after a while or we accept them from other people simply as social stereotypes. In the probability, therefore, there are reflected the objectively possible (because known from experience) 'from-here-to-there' conditions of the frames of reference of the world which surrounds us, assails us or responds to us. The effect of a particular reaction lasts 'from here to there'. Within this we sum up in the space of probability the kind of special effect with which it will assail us as individuals. Ágnes Heller has an apt description in this regard: "To act in a sphere lower than the space of probability, in that of mere possibility, would inevitably make orientation and practical successfulness hopeless, or at least it would reduce them to such a small percentage that man would at once become incapable of living. The situation would be the same if man tried to act in a higher sphere, in that of scientific safety. Under the circumstance he would have to check the content of scientific safety before each step. For example, he could not switch on the lights till he had a complete knowledge of electricity." [Ágnes Heller: *Társadalmi szerep és előítélet* (Social role and prejudice), Budapest 1966, p. 80.] At the same time, thinking in terms of probability encourages not only the correct action but also the creation of illusions. In our daily life we attach more importance to the verisimilar than to the real. For indeed, something may yet issue from the former, whereas the latter is of secondary importance from a teleological point of view since it represents only an alternative condition. That gossip, for example, always seems to be more authentic than concrete facts is due to its looking more obvious, though the basis of its acceptance is the same logic of probability as we find in the case of quick action.

Let us come back, however, to the ontological structure of object-making and objectivity. As we have seen, in object-making we do not identify ourselves with the totality of 'things in themselves' but we merely use them in one or more of their relationships, while we also identify this relational existence with the things themselves. Objectivity in the ontological sense is therefore a greater totality. With reference to Nicolai Hartmann we might even say that it is a 'transobjective objectivity': only that which is capable of being 'grasped' and made as well as cognized rates as an object for immediate praxis. This configuration of being which is beyond this relational existence and which maintains this determinateness of form becomes 'transobjective' *('übergegenständlich')*. Object-making is unable to confront the totality of things. It cannot encompass this transobjective objectivity in terms of either practice or epistemology. At the same time 'used' things are not only 'less' but also 'more' than the ontological totality, namely praxis and mainly conscious apprehension also add to this relational existence as we grasp these things, complementing them with characteristics which ontologically they do not have. We grasp the things 'as such', but this 'as such' may also include some subjective, fantastic and nonexisting elements. The deceived person treats loyalty, honesty or imitation as something genuine because it was given to him 'as such', and he not only accepted this analogous form but he may also have embellished it. Ontologically we are confronted therefore with an accumulated structure. Genuine objectivity is a totality the extension of which is perhaps unknown to us, a totality of which object-making grasps only a segment of relation, albeit objective and successful practice requires that the relational existence we use bring the totality of things in motion. The thing that we grasp as an object will simultaneously call into action that larger unit of which it is a projection. Practical consciousness fails to perceive this mediation. The accumulated system of objectivity simply does not exist for it: it can realize and grasp only the objectified relationship with which it identifies the transobjective one about which it may have only vague notions. In analyzing the illusion of the sources of income, for example, Marx says that the individual configurations of income (profit, allowance, wages) get separated from the uniform process of the production of surplus value as different and independent forms, but as soon as they pick up at all particular figures which are indifferent to and independent of one another and which are regulated by different laws, their common unity, i.e. surplus value, and consequently the nature of this common unity, becomes increasingly unrecognizable and does not manifest itself in the appearance. But one reason of this concealment is to be sought in the fact that each particular source of income is judged according to the manner in which it has been acquired, i.e. as it has been made an object. They are not judged in their original unity, i.e. in the totality of their transobjective objectivity, but as the individual groups identify themselves with these forms. These forms as such practically determine the activity of the individual capitalists etc., and supply their motives as they also reflect as such in their consciousness. [Cf. *Surplus Value,* Vol. III.] The reason why the dialectic of the two different objectivities (ontological totality and functional existence) is not perceptible is that in the course of object-making everybody focusses on and obeys that which is important for him, and by this means he will himself conceal the mediation which links the functional mode of existence with the entirety of things and with the social system of functions.

Objectivity is brought about socially, but only through the capitalist production of commodities will this preformed figure become of average character. The original shape of things is covered up by the average form which can be used by all and sundry. It is this average form which I grasp in object-making and which I continue to improve and, in the meantime, I also adapt myself to the average habits, expectations and standards. The only way in which the average object becomes an accepted, functioning objectivity is if the average person handles it with an averaged outlook and practice. Remember the clichés of the modern ideal of beauty. It is a truism that fashion's annually changing demands for new forms shapes and standardizes women's features with a necessity affecting, as it were, the physiology of their faces and their bodies. Meanwhile, an average standard of beauty is called forth, to be worn and then replaced in the same way as some stylish dress. But this stereotyped, average standard of beauty owes its rise to the circumstance that the features of some woman who is prettier than the average are fashioned by the expectations of her friends and admirers around her. The new 'mask' is made 'for them' but with the singular objective of scoring success with respect to each of them. It is this attitude — trying to fall in with 'everybody' — which will make average not only the features but also the personal habits, but only because the expectations of the 'others' are guided *a priori* by this average standard. What was the uniqueness of the face becomes the average beauty of average standard, but only in relation to the average requirements can it function as 'beautiful'.

It seems as if the socialness of object-making, i.e. compliance with the various individual requirements, makes objects and people average as if the average were merely one form of social generality. For indeed, working tools, for example, develop by passing from hand to hand and, in the process, the individual handiness increasingly wears off them, leaving only what renders them usable for everybody. (Marx notes, for example, that in the age of pre-machine manufacture there were some fifty different kinds of hammers in circulation, and that by the time the organization of work had been established in the factories with the standardization of individual and particular working conditions, there were only about a dozen of them left.) It is undeniable that usage, the socialization of object-making 'polishes' the means of production, but this process is still not the same as averaging. Each collectively used tool — thus, for example, language too — is the product of this 'common shaping' through which it becomes of universal use. We adjust the objects to the logic of the working process. Without it, neither the social organization of work nor the improvement of the tricks of the trade or their communication to others, nor the utilization of new discoveries would be conceivable. The point is, however, that it is not the individual who adjusts himself to the requirements of the objects but it is exactly the other way around: it is we who adapt the objects to the requirements of the social individual. Standardization on the other hand moulds the individual's requirements, his way of looking at things and his expectations to the stereotyped objects and forms which regulate social intercourse. The dictate of fashion, for example, means that everyone has to wear identical shoes or cut of clothes. To the individuals it is immaterial whether the clothes they wear 'suit them' personally and bring out the favourable or unfavourable qualities of their physical attributes or traits of character. They are guided by the fact that everybody likes that particular style. And the more they hide the uniqueness of their individuality the more they can count on success.

The most important characteristic of the average lies in that it hides the individual features of people, and that it renders objects visible and usable only insofar as they promote the successful exchange of actions. The objects of prestige consumption, for example, are endowed with a plus value, resulting not from their original features but from the anticipation of the impression they will make on others. Their role is to ensure success and to arouse envy and admiration. It is also true, however, that they may be of inferior quality, less suitable for use than objects without this plus value. But this objective flaw cannot be seen as it is concealed by the possibility of enhancing their possessor's prestige.

This is how both the unity and the difference of average and genuine objectivities come into being. The 'average' conceals the real things which convey it. It functions as an illusion but more can be achieved in the social exchange of actions with its aid than with the 'things themselves'. The individual habit of people, the real structure of objects will be the vehicle of this average 'mask' which settles upon them in the course of social intercourse. But this adopted form of presentation also has an objectivity of its own just as the 'original' objectivity has. Its origination and rule are conditional on the universal equivalent which controls commodity production. The individuality of people, the uniqueness of the objects, must first be reduced to the condition of exchangeability, i.e. to the formula of the universal equivalent, before they can get into the bloodstream of society. Furthermore, they can figure as social entities, as objects existing also for others, only insofar as they avail themselves of this equivalent form. Though this 'averaging' tendency had been present in commodity production from the very beginning, it was only in capitalism that it became a universal social regulator, a model which can bring under its direction all the phenomena of life. And here we are really dealing with a model. The process of producing average price and average profit affects — in mediated forms — all aspects of life, allowing the exchange of actions to materialize everywhere only on an 'average level'. And in this economic process the average is not a datum of the mental view but it is a given attribute which is created subsequently by the interaction between the individual producers and which also determines the further course of their production. Its existence is corroborated by the fact that a departure from the average may ruin individual people, though its requirements cannot be foreknown since they develop only *ex post facto,* in the struggles of the market and the competition, as the equalization of the goods produced on many different levels of value.

With the rise of the 'average' a social regulator has come into being which can govern all levels of social life more effectively than any other earlier forms of control could. This difference is obvious when we compare the different types of regulators that have developed in the course of history. In classical antiquity and to some extent in the Middle Ages it was the example, the precedent, which served as a guide for individual decisions. The gesture of the heroes of Thermopylae had for a long time served as an example for Greek *polis* ethic, for the moral standard of the community, and shaped individuality. But even at the beginning of the development of Christianity this exemplariness constituted an effective means of social control. The legend of Christ, the exemplariness of the saints, represented a normative power capable of directing the individual's mode of life. One can relate to an example, however, only in an individual manner, but having once submitted to it one also becomes a member of the community. One can see in the example virtually the fulfillment of one's own life, and either one will attempt to achieve this — emotionally already experienced — requirement in reality, too, or content oneself with being mentally associated with the example while only conforming to its demands externally. The former is characteristic of *polis* ethic, whereas the latter is engendered by the attitude of disintegration of the religious ethos of the Middle Ages. Parallel with this disintegration there came into being traditions or legends, no longer summarizing the requirements of communal existence in the suggestive effect of individual human accomplishments but comprehending a more abstract system of standards, the abstract instructions of the 'what shall we do' abstracted from the particular deeds, and handing them down from one generation to the next. Here, too, the standard is given in advance, for indeed everybody knows the difference between right and wrong and consequently what he must or must not do. The social control is achieved by members of the community representing the assertion of these standards towards one another. The guardians of tradition enforce the observance of these requirements by virtue of their sheer presence, power or authority. This type, therefore, became a universal regulator in socially conservative, identical and, so to speak, nearly self-reproductive formations which tried to preserve the existing social relations, particularly in the 'Asiatic mode of production', in the stagnating phases of feudalism or in village communities. The way it takes effect has, however, nothing to do with averageness. Though the safeguarding of traditions will standardize things (cf., for example, the centuries old immobility of the peasant way of life), it lacks nevertheless the most important feature of averageness: the subsequent equalization of the many, individually different motives and life-styles.

The heroic period of bourgeois society creates a new kind of regulator, namely the system of requirements of ideals. Tradition still influenced the actual present of people as a historical reality and demanded the reproduction of the past. Revolutionary ideals confronted people with a prospective requirement and at the same time also abstracted from the everyday reality of their life. Their life was redoubled: consciously they lived in this ideal system of norms, while in practice they were incapable of satisfying their requirements. Hence 'guilty conscience' became the regulative power, i.e. the daily experience that the individual person is incapable of living 'in an ideal way' and that it is only through a permanent guilty conscience that he can realize his compromises and the materiality of his 'bourgeois' way of living.

In the consolidated bourgeois order ideals tend to lose more and more their regulatory functions and are replaced by 'the average' which has a stronger influence because it develops its standards after the event as a material demand, as the average of the many different kinds of individual and contingent decisions, making the already existing tendency a standard to be followed. The average imposes more real demands on the individual person who by complying with it has a greater possibility of achieving success, whereas deviation from it will entail more serious consequences. Simultaneously, however, it leaves the semblance of individual extravagance and freedom.[10]

This is when the 'object itself' separates from its average and functional existence, although ontologically these two forms of objectivization are based on each other. The objectivity of illusions originates in the difference and the hidden identicalness. For indeed, the individual can handle the average function of the objects as reality, but he is unable to see the structure whereby 'secondary objectivizations' created by social conditions are superposed upon the original objective data. There occurs a 'pseudo-concrete' objectivity in the reflection: the clichés, the stereotype forms and the attendant and comprehending 'panels' of thought hide the individual features of things because they can be handled, objectified and exchanged only in their 'clichéhood'. And since this average form can be used more successfully and since it is more prevalent, an attitude will develop which will regard this secondary objectivization as the sole form of objectivity.

Hegel was the first to describe this kind of transformation of objectivity, the phenomenon of this secondary objectivization. For him the 'thing itself' *('die Sache selbst')* means the socially useful and interchangeable objectivity which exists for the other but which differs from itself. (He observes essentially the dialectic of exchange value, the process that relates to the equivalent.) In the *Phenomenology of Spirit* he writes: The essential thing *('die Sache selbst')* is present as a thing in itself or as the reflection of consciousness tending to itself; but the supersession of the elements by one another is manifest in it by being posited in con-

[10] "This accidental character as such is only engendered and developed by competition and the struggle of individuals among themselves. Thus, in imagination, individuals seem freer under the dominance of the bourgeoisie than before, because their conditions of life seem accidental; in reality, of course, they are less free, because they are to a greater extent governed by material forces." [K. Marx and F. Engels: 'The German Ideology', in: K. Marx and F. Engels: *Collected Works,* transl. C. Dutt, W. Lough and C. P. Magill, London 1976, Vol. 5, pp. 78—79.]

sciousness not according to their being in themselves but only for a different consciousness. It turns one element of the content outwards and shows it for others. According to Hegel the first step is that the value in use, the configuration 'being for others', becomes independent and thereby hides the 'being in themselves' of things. But he knows that this form of presentation reacts also upon the object itself but this intrinsic dialectic is not perceptible. As he writes: It does not contain a single element which consciousness turns only outwards and another which it holds back only in the inside, but it keeps alternating them because it must render both the one and the other an essential element for itself and for the others. *[Ibid.]* But the second step is already the particular social transformation of things or objects. It is the averagely developed objectivity which is refined in the course of the making an object of each and every participant to become 'just as it is', i.e. an average form which differs from its original nature. Hegel continues his chain of thought: The fact is neither only the thing nor an activity which runs counter to being (to that of the thing — M. A.), but it is an essence whose being is the being of the activity of the single individual and of all other individuals, and whose activity is immediately for others, in other words it is a thing and it is a thing only as the activity of all and each. *[Ibid.]* The average form of objectivity takes shape in the relation between the total social activity and the individual activity. But Hegel also stresses the importance of the dialectical structure whereby this social objectivity conceals the real content of things: the average manifests itself in consciousness as an illusory object. To begin with, writes Hegel, what matters to consciousness is not the thing as an object but as its own thing. *[Ibid.]* The 'usability' of the average configuration, the mediation of ownership conceals the difference between the universality of the objectified thing and its individual content. One deceives not only others but also oneself by this means. Because in the course of this collective transaction there occurs an interplay of personalities in which they deceive themselves and also one another and see both as being deceived. *[Ibid.]* The irony of objectivity is, however, that in being deceived by each other they are nevertheless both creating and being punished by a law of which in their preoccupation with the 'objects themselves' they have no knowledge at all. The illusion takes shape in the averaging process of social reception and further transmission and is finally registered in the limitedness of conscious 'recognition'. But the rub is that this form of illusion is a cunning phenomenon. Society realizes itself in it. The falsely known objectivity is at the same time a precondition of the genuine movements.

The problem of investigating *'die Sache selbst'* surfaces again with Husserl but in an inverse relation as to its content.[11] While Hegel understood clearly not only the process of this configuration (he treated it as a form of consciousness that is characteristic of capitalism) but also the connection between its known and recognized falseness and hidden truth, Husserl proceeds only from the result, from the validity, from what is subjectively known. This is the basis on which Husserl wants to reconstruct the world of 'phenomena', a world which is ultimately subjectivistic and illusory. He writes that not a single form of presentation can arise as an absolute datum since the 'known thing' *('Gewissene')* can figure in the sphere of practical interests only as the norm of a certain advantage. [Cf. Husserl: *Ideen zu einer reinen Phänomenologie und phänomenologischen Philosophie,* Halle 1913, p. 80.] The observation that the form of presentation is not the same as the thing itself is obviously right but, instead of tracing back the objective transposals and the heart of reality of this mediate and averaged form, Husserl subjectivizes this intrinsic truth: it is consciousness which lends meaning to things. He writes that here reality corresponds to certain effective units of value which by virtue of its essence absolute consciousness uses as a certain relationship which imparts and also possesses some meaning. [*Ibid.,* p. 106.] In our everyday praxis we do indeed encounter some 'reality-creating' element like that but only as an illusory (because immediate) performance in which the relationships concealed for the individual acquire a distorted form of expression and which, precisely because they also carry the elements of actual content, we can use as valid forms. It is in this way that we can handle successfully our objects in a false illusion. Husserl, however, generalizes the habits accepted in the immediacy of everyday praxis: the 'thing itself' is identical with the meaning assigned to it by consciousness. That is how teleologically transformed things, i.e., those which we have understood, assume a falsely ontologized form.

[11] Both Husserl and Scheler made liberal use of this Hegelian term. Scheler writes, for example, that phenomenological experience is capable of supplying the facts themselves and therefore of immediately grasping them and so not by the mediation of some kind of a symbol, sign or reference. Thus we can say that in principle any non-phenomenological experience can happen only through some symbol, and that it is therefore only a mediate experience and never the thing itself. Only phenomenological experience is symbolic in principle and consequently only this is capable of filling out every symbol. [Cf. M. Scheler: *Der Formalismus in der Ethik und die materiale Wertethik,* Halle 1929, p. 46.] Of course, the 'thing itself' manifests itself from the outset in its immediate, socially given and consequently average form. The condition of its symbolicalness is that it simply takes the socially given symbol (cliché, equivalent form) for its basis.

Things can exist only through our intentions and as the objects of our aims. The objective forms of their substantive nature are created by a 'collective intentionality'. With this, however, we will lose sight of the connection between objectivity and the socially formed 'secondary objectivity', i.e., the hierarchy of and the difference between the natural and the social modes of existence. Not only will the world become subjective but only illusion will seem real in the resulting objectivity. It even becomes existent by virtue of the validity of the 'known' configuration. As the phenomenological method can bring itself into due relations to the surrounding facts only in objectivity: the specific characteristic of phenomenology lies in the fact that within the sphere of its eidetic generality it encompasses all perceptions and sciences according to such definitions as are immediately discernible in them. The range of our investigations includes the import and the legitimacy of every possible immediate point of departure and immediate step. [*Ibid.,* p. 118.] It is therefore impossible to separate illusions from existent phenomena on such grounds. Husserl perhaps has not even the intention of taking these two presentational modes to be the same; on the contrary, this is precisely what he tries to avoid in his later development (primarily in the *Krisis*). On the other hand, by subjectivizing objectivity he leaves this possibility open: things may be illusions or genuine connections. There is no possible way of effecting a clear-cut separation between the two. This methodological flaw is particularly obvious when he writes about the eliminability of illusions: If our perceptions are false it means only that they are confronted with those further percepts of ours which show with certainty what the true relationship and what the illusory one is. [*Ibid.,* Vol. 3, p. 417.] Husserl does not even surmise that illusions may also have a prolonged objectivity which cannot be eliminated by further experiences and which is incapable of being transcended in immediacy.

That is why Max Scheler suggests that the point of departure for the phenomenological method would be simply to eliminate the differentiation between the 'real' and the 'illusory'. The 'phenomenon', says Scheler, is indifferent with regard to the differences between illusion and appearance. Reality dissolves in the sphere of effective, average objectivity. Its articulation, mediatedness, essential positedness and the process of its object-making are lost to view. [Cf. M. Scheler: *Op. cit.,* p. 43.]

Let us once more refer to Adorno, who observes correctly that behind Husserl's objectivity there lie the averageness of the commodity form and its form of presentation and that he takes its immediacy for 'the' reality: The equivalent and the model of the reification of logic as the self-alienation of thought is the reification of that to which the thought relates: the unity of objects hardens for thinking into an identity so that consciousness is unable to perceive the changing contents of this identity and is capable of registering only the bare form of this unity. This unity suggests the commodity form whose identity resides in the 'equivalent' of the exchange value, and thereby in false consciousness, in an in itself baffling social relation, the subject. [Cf. Th. W. Adorno: *Zur Metakritik der Erkenntnistheorie, Op. cit.,* p. 79.] While in Hegel, in the *Phenomenology of Spirit,* this 'thing itself' deduced from the commodity form is explained by analyzing the difference between the illusory and the real stratifications, in phenomenology it is precisely the exclusive reality of the commodity form which becomes the model. Phenomenology, on this basis, can dispel reality and render the category of appearances inapprehensible.

Let us once again examine the elementary sphere of the origin of illusions. As we have seen, the false configurations of presentation emerge in their objective quality (with a relative objectivity, of course) for certain 'positions'. The configuration which manifests itself in the relation of 'being for others', i.e., for an individual acting in either a social or a natural situation, acquires meaning by virtue of analogical reasoning. In other words, illusion becomes a significant image in its consciously interpreted quality. The position thus determines the mode of appearing (which we call functional presentation) or, in other words, the elementary configuration of illusions. Human praxis continues this objectifying sequence. It is due to object-making that the things that people use and 'make' will acquire the quality and the form of presentation in which they manifest themselves for consciousness. The false form of presentation is vindicated in practice. The virtual form of presentation created originally in the mutual relation of position and function ossifies in the course of object-making. Illusion becomes a category of praxis, of everyday practice, and it can be ended only by transcending immediacy, after leaving the sphere of individual practice. Hence the ultimate philosophical condition of illusions is the relevance of immediacy, or in other words that consciousness and practical relation reach a particular group of appearances only through a single way of approach. Illusions are engendered for immediacy, whose sphere establishes them as 'false realities' or, for the ordinary person, as natural ones. The difficulties and at times the impossibility of transcending immediacy are developed by the structure of capitalism and then by manipulation.

As we have seen, an ontological precondition of the objectification of illusions is subjectivity, i.e., the conscious, practical work of the acting individual. Therefore, in the next chapter — after briefly reviewing the relevant part of the history of philosophy — we shall examine the subjective conditions of the formation of illusions or how illusions are elaborated by practical consciousness. First of all we shall deal with its characteristic feature which results from teleological activity and shows reality in an inverted relation and scale of values. We shall proceed then to examine the interpretative structure of ordinary consciousness, the first phase of which is the formalization of relationships. It is by this means that practical consciousness renders similar the in reality dissimilar objects and relationships.

2.4. THE WATERSHED OF MODERN PHILOSOPHY

It was not the twentieth century that gave immediacy its place in the forefront of philosophy, nor was it discovered by neopositivism or phenomenology. It is a phenomenon which influenced the entire intellectual development of modern times. We suppose that nearly all questions of this development have stemmed from the crisis which was represented by the process in which the sphere of immediacy achieved independence. The rise of particularity, the first eruption of the difference between the individual sphere of life and the total social relations, was a concomitant phenomenon of the birth and the development of bourgeois society, while in terms of thought it brought in its train a profound spiritual crisis and two attempted solutions which at times approached hysterical forms. On the one hand, it brought with it not only the discovery that reality had disappeared and that its appearance which could be experienced immediately both outwardly and inwardly was unreal but also the theoretical basis for this process of 'derealization'. (This attempted solution led to the emergence of subjective idealism and then to the abstract and subjectivistic forms of rationalism.) Hence from this point of view it seems as if the empirical sphere of phenomena which can be seen and experienced were purely the work of the subject, as if the objectivity of the world had melted away; as if only the great 'Whole' existed in reality and its attributes and accidents were no more than illusory and subjectivistic notions. (The failure of the British Enlightenment to maintain its foothold, the Berkeley-Hume line, might stand for the branch which represented the unreality of immediate appearances and their subjectivistic nature, whereas the Descartes–Spinoza–Fichte line led to a variety of attempts to rescue the 'Whole', the substance, at the cost, however, of the 'unreality', in various degrees, of individuality.) The other attempt at solving this crisis runs an opposite course: only that is real which exists immediately, which is tangible and which manifests itself for the individual. Here the eruption of the difference between immediacy and the total motion led to the attempt at rescuing from this crisis at least the reality of the particular facts and the stability of the individual sphere of life. (This attempted solution is characteristic of empiricism, sensualism, and moreover early materialism.) Ultimately even the emergence of modern idealism can be understood as an attempt to bridge this crisis. The 'more real' spiritual totality (idea, ego, *anima mundi*), whether in a subjectivized or an objective form, reflected the fact that at the birth of modern society the total movement was 'more real' than the immediate facts, events and phenomena, but this totality could be experienced and reflected only conceptually. Hence the problem pervading nearly every stage of the modern development of philosophy is to make a judicious decision as to the quality of immediacy: to prove whether it is illusory or real. More exactly, all philosophical problems become steeped in the atmosphere of an intellectual struggle with illusoriness.

The crisis, the change in the outlook, is first revealed perhaps in the transformation of the conceptual interpretation of the problem of the whole and the part. For Descartes it is still the unproblematic whole that represents the natural criterion of truth and reality, and in comparison the parts are either imperfect or simply false.[12] So Descartes already suspects that there are 'true' parts as well as false ones, that the composition of a world made up of accidents is not inarticulate. Yet he still ascribes articulation to cognition and still cannot have any knowledge of its ontological structure. But the flight to the Whole is a critical search for some basis in his case, too. For if we compare this idea with Hegel's self-confident concept of totality ('only the Whole is true') and the resulting historical and ontological construction of the world, we realize that what appears in Descartes' case is still only the disconcerting state of agitation aroused by the question. Out of this crisis there arises the recourse to rationalism. Descartes already suspects that man's chief anxiety is to

[12] "I do not dispute that a greater perfection resides in the entire nature of things if some of its parts are free from error while others are not, whereas perfection would also be less if all of its parts were identical with one another." [Verbatim translation. Descartes: *Untersuchungen über den Grundlagen der Philosophie,* Heidelberg n.d., p. 53.]

create a meaningful relationship between illusions and reality, but he still regards this as simply a matter for rational adjustment: the delusions of the sensory organs must be excluded. [Descartes: *Op. cit.,* p. 105.] For all that, rationalism is more than merely the elimination of sensory immediacy. It is also the positing of an abstract spiritual totality (God), which we must keep to the fore in order to succeed in getting rid of the deceptive illusions of the 'parts'. The twofold character of reality, the particular truth of immediacy which may even conflict with the totality while it is its form of presentation, is still an unknown relationship for Descartes. Only the rejection of immediacy is implicit in his critique of empiricism. Its totality-building role disappears.[13] The cutting edge of the critique attacks a genuine problem but the solution hides the true relationship. There is still no sign here of the objective obstacle to conceptual comprehension. Error seems like the relation between the receiver and the received. Man wants to see more in things than his knowledge of facts allows him to and he projects this intentional surplus on to the things. [*Ibid.,* p. 84.]

The Whole, the movement of totality, still depends here on the faulty or normal function of the positing work of cognition. But even Descartes is affected by the crisis experience which the structural change of totality brought about. The British Enlightenment, at the opposite pole of the development and as the devil's advocate of 18th-century thought, launches a direct attack against the category of the whole and the part. On the grounds of vindicating immediacy it regards this concept as untenable. It chooses the common sense of everyday life, the essential requirement of the bourgeois mode of life, as the criterion of thought. What exists is what we do and what we work with and the whole is only a fiction. The discovery of the gap between primary and secondary qualities (introduced by Locke into philosophy) explodes the until then intact category of totality. Locke grasps the secondary qualities, i.e. the perceptible ones, and not only from a subjective point of view; he even denies agnostically their deducibility from primary qualities,[14] sanctioning

[13] "For it is undeniable that those elements of reality which offer substances contain for me more and, so to speak, a higher order of objective reality than those which represent only conditions and accidents." [*Ibid.,* p. 53.]

[14] "Besides this ignorance of the primary qualities of the insensible parts of bodies, on which depend all their secondary qualities, there is yet another and more incurable part of ignorance, which sets us more remote from a certain knowledge of the co-existence or unco-existence (if I may say so) of different ideas in the same subject; and that is, that there is no discoverable connexion between any secondary quality and those primary qualities which it depends on." [J. Locke: *An Essay concerning Human Understanding,* Book IV, Chapter III, Paragraph 12.]

hereby the redoubling of reality in philosophical terms as well. There arises an odd solution. The major proportion of the immediately perceptible data acquire an uncertain and arbitrary mode of existence. Their illusion-like nature, or verisimilitude is parenthesized. Furthermore, the road is open for a consistently idealistic conception in which the homogenization of the Whole reconstructs the unity of reality in the process of primary qualities, i.e., in terms of concepts such as ideal or abstract size, number, motion etc.

This conception grasps the data of bourgeois routine in their real untruth.

Of course, this viewpoint also simplifies the category of immediacy subjectivistically and epistemologically, absolutizing it in accordance with individual experience and sense-perception and not concerning itself with its objective origin. Behind the conception there is the arbitrariness of the original accumulation, the individual and unique phenomenon of careers, the practical plasticity of reality and its bilingualism. All this is naturally wrapped up in the tedium of Locke's formulation.

The basic categories of practical consciousness, of everyday praxis in the bourgeois sense, are also formulated here for the first time, i.e. considerations based on probability and not according to the *a priori,* the qualities of the form of consciousness which thinks in terms of analogies, the phenomenon of mutually contradictory degrees of knowledge; in other words, all those phenomena which have become historical facts of the sphere of immediacy and the total movement which is losing contact with it. The theory of primary and secondary qualities is only the ultimate philosophical abstraction of this convenient viewpoint which originates in common sense. But no sooner had this duality been enunciated than a possibility arose for a new kind of homogenization: the possibility for a consistently idealistic solution. These consequences are epitomized in a hysterical form by Berkeley and in a carefully weighed, seminal system by Hume. This is how immediacy as a historical experience of crisis is first formulated as subjective idealism.

Already Berkeley realizes that immediacy is a deceptive datum. It is an existing and a non-existing configuration. That is why he argues against scepticism where no clear-cut decision has yet been made between apparency and the verisimilitude of immediacy. Berkeley homogenizes this category, trying to eliminate the charge of illusoriness brought against the daily life of the bourgeoisie. But the only way he can do this is to definitely subjectivize this category: "Nothing seems of more importance towards erecting a firm system of sound and real knowledge, which may be proof against the assaults of scepticism than to lay the beginning in a distinct explication of what is meant by THING, REALITY, EXISTENCE," he says programmatically, and then goes on: "If they are looked on as notes or images, referred to things or archetypes existing without the mind then we are all involved in scepticism. We see only the appearances and not the real qualities of things. What may be the extension, figure or motion of anything really and absolutely, or in itself, it is possible for us to know, but only the proportion or relation they bear to our senses." [Berkeley: *The Principles of Human Knowledge,* 1710, Paragraphs 87 and 89.] In my opinion we are at the source of subjective idealism, a source which is nothing else but the bourgeois vindication of the homogeneity of immediacy. Eighteenth-century philosophy came into collision with the contradictory layers of immediacy, with the fact of recognizing that there was a difference between the total movement and the superficial phenomena, and since the primary need was to maintain the reality and the homogeneity of immediacy and also to ward off the charge of illusoriness, the first obvious solution was to subjectivize this sphere. Theoretically this meant that the individual sphere of life, the world of phenomena existing for the particular man, had to be the sole sphere of reality. This is how the world had become homogenized. The contradiction which had still played a role with Locke vanished with the disappearance of the contradiction between 'being for us' and what manifests itself 'otherwise' in the total movement. But this solution also meant that the world had to be narrowed down to the practical domain of sense-perception and experience, i.e. to the everyday substance of life.

And indeed, on this basis the contrast already disappears between the illusory and the real, and Descartes' dilemma — i.e. to what extent the total movement is inconsistent with partial appearance, the substance with the accident — cannot even arise. This bourgeois and ideological intention is distinctly manifest in Hume's case who in his writings on political economy clearly recognizes the role of the double reality of money (gold), but ascribes a genuine (real) existence to the one and a quasi-existence to the other (to its social function).[15] This is the conceptual ground on which we find the endeavour to homogenize the

[15] "It is indeed evident, that money is nothing but the representation of labour and commodities, and serves only as a method of rating or estimating them." [D. Hume: 'Of Money', 1752, in: *David Hume's Writings on Economics,* London 1955, p. 37.] Marx, too, refers to him as a sober economist. Still, he deduces the ideological consequences of his economic observations in positing a quasi-reality.

twofold character of things. But the only way in which this defensive struggle against illusoriness can meet with success is if it subjectivizes the sphere of immediacy.

The reason why the conceptual crisis as to understanding immediacy erupted in England was due to the circumstance that it was in that country where the twofold aspect of this sphere first manifested itself. In the age of the primitive accumulation of capital and of the industrial revolution it seems as if each and every citizen is given an endless series of immediate possibilities, while in practice it turns out that the reality of these possibilities is, after all, extraordinarily limited. Hence, in view of these facts, is this sphere of possibilities real or only illusory? The facts of daily life always vibrate in a dual light. More and more elements look empty and simultaneously — in individual practice — real. However, in this victorious period the ideological mission of philosophy is to surmount this crisis. It wants to take notice of only what is real in this practical relation. The measure of apparency seems to be of no importance for the time being and hence the fierce struggle against scepticism, hence the nearly hysterical flight to subjectivized reality.

In view of this, however, we must also reshape our traditional idea of subjective idealism. It is not simply the reactionary thesis about the hiding of reality but it is the first ideological conclusion to be drawn from the practice of the primitive accumulation and of the industrial revolution. The second wave of the British Enlightenment has no intention of continuing to analyze the reality of the age's experience, as Bacon did, because it suspects that it may reach a problematic final result. For the same reason it enthusiastically consolidates the immediate data into a single reality, while alarmedly protests against any doubts or pretensions to truth which attempt to gain a deeper insight of the relevancies and consequently also against any investigation of the contradictory relationship between immediacy and the total movement.

The crisis of the transformation of the sphere of immediacy plays a decisive role in both the emergence of agnosticism and the birth of modern idealism. In the history of philosophy it is a cliché that Hume influenced Kant and that Hegel, the dominant personality of German idealism, was in close association with the theories of the British economic model. Notwithstanding these influences, the German philosophical development pursues different solutions. The backward state of the country also meant a certain distance with respect to this complex 'phenomenon'. Thus classical German idealism, although we conjecture that it also receives impulses from the crisis of the sphere of immediacy, still has the capacity to construct a bridge between the two extremes of the 18th century and to find the dialectical solution leading to the new synthesis of immediacy, illusion, appearance and the total movement. For indeed, the development from Kant to Hegel at nearly every stage seeks the answer to the question of how it might be possible to overcome the false alternative according to which immediacy (empiricism, everyday life) is nonexistent or negligible as compared with the Whole, or else that only immediacy, the bare reality of experience exists and that any other reference beyond this is mysticism. Likewise, it is also well known that Kant is in thrall to this double task. The fundamental contradiction in his system — the gap between theoretical and practical reason — gives eloquent proof of the great philosophical mission of the 19th century and of the magnitude of the task. It is Kant who first tries to formulate the correct answer through the theory of antinomies and the concept of transcendental illusion, but it is the young Schelling who gives immediacy as a question an explicit form. With him it comes up as a crude crisis problem and it is only with Hegel that it will finally come to a dialectical rest.

It would seem appropriate to start our outline of the problem with Schelling. His oeuvre itself (his entire philosophy) is the manifestation of a permanent crisis, not only through its constant veerings but also as a result of his gradual shifting to the right in the course of which he was to drift from his *Naturphilosophie* (combining the elements of materialism with dialectical studies) to religious mysticism. In the papers of his Jena period, when with Fichte he tries to elaborate the theory of the new structure of reality, there arises time and again the question of the twofold aspect of immediacy and the need to elucidate it. His point of departure is already a plan for homogenization but he labours to reformulate and modernize the Cartesian category of the Whole, in other words to encompass philosophically the entirety of the bourgeois order. In his system the facts of the immediate world are once again connected with the total movement (substance, God, the Whole). This reconnection is, of course, unstable for the time being and is full of mystical elements, but there is one aspect in it which is progressive by all accounts: he tries even unconsciously to restore the organic nature of the entirety of social praxis, the incongruous connection between the total social processes and the individual sphere of experience, in other words, the very same relevancy which was lost in the line of the British development. In his view immediacy is not an 'imperfect' or illusory datum (as in rationalism), nor is it a single, subjectivistic basis for reality (as with the

British), but he takes note of the double aspect of this category and attempts to construct his system upon it. The relative nonexistence of the particular in relation to the whole is the concrete existence of reality itself, writes Schelling, and by recourse to this relationship he simply pushes aside the until then predominant false alternative. [Cf. Schelling: 'System der gesamten Philosophie', handschriftlicher Nachlass, in: *Schellings Werke,* München 1965, Bd. 6, p. 118.] The conceptual invention is strikingly modern. On the one hand, he regards immediacy as a particularity, and by this means he succeeds in resolving the more dialectical condition of relating to the whole, still lacking in the earlier attempts. On the other hand, he also realizes that the fact that the being of things for us and their relevance to the whole differ from one another has actually no bearing on their existence, for indeed concrete being (i.e. 'being for us') is conditional upon the 'voidness' of the relevance to the whole. This is how he solves — at least in its germ — the fundamental question pertaining to the mode of existence of immediacy, i.e. the dialectic of the relationship and the contradiction between the outward forms of presentation and their intrinsic hidden content.

Schelling's achievement was that he conceived immediacy as objective and phenomenal. He did not see it as a phase of cognitive faculty, as not only British sensualism but even Kant saw it despite all his brilliant endeavours. Schelling handled it as an autonomous existent and also as a form of relevance. He suspected that there is a frame of reference in immediacy, by virtue of which things appear to us otherwise than they do for the Whole. Of course, not even he could find the answer as to what sort of ontological mediating process creates this differentiating condition, and presumably it was due to this ontological uncertainty that he was later attracted to irrationalism. However, no doubt it was due precisely to his recognition of the objective structure of immediacy that he created the system of objective idealism, in opposition to a philosophical tradition which, sustained partly by British and partly by Kantian and Fichtean sources, seemed to be nearly irrefutable. (We can see here, too, how much the conceptual elimination of immediacy is a key issue in the philosophical trends of modern times.)

He cannot, however, resolve the question of the double content of the category of immediacy. He suspects that it has an objective structure and that it has its roots in reality, but he is unable to comprehend the particular element of immediacy which is implied in the conceptual and cognitive performance as the other aspect of appearance. This weakness is due to his own conceptual apparatus, to the 'intellect-centred outlook' which resolves the 'spontaneous process' of reality (i.e. its ontological transformation) and the manner of cognition in a single process. No doubt, the solution points towards irrationalism. However, not even this element is so unambiguous when one examines the development of the problem. In his philosophy of nature this (episto-onto) unity still figures in the sense of a 'phenomenon', as the unity of the objective mode of manifestation of reality and the motion of the conceptual apprehension as it proceeds from immediacy to mediatedness. For example, he writes that the product of perception is not in the least some sort of original thing, but is the joint product of objective and subjective activities. [Cf. Schelling: *Ideen zu einer Philosophie der Natur,* Leipzig 1797, p. 142.] In other words, the notions that take shape in perception are the joint modes of existence of the active individual and of the reacting and resisting external world which is taking shape in the meantime. But the snag is that Schelling absolutizes this cognitive phase. Instead of recognizing it as one of the elements in a step onward, an element from which it might be possible to disentangle the difference between the ontological structure and the subjective reflection, he expects to find the ultimate end of knowledge in it. [*Ibid.,* p. 140.] Later on, the category of the 'intellect-centred outlook' will unambiguously 'produce' reality.

As regards the system of objective idealism, these elements indicate merely the boundaries. The young Schelling's intellectual maturity can be felt when he grasps the sphere of immediacy as an objective relational existence. In his manuscript to which we referred above, *System der gesamten Philosophie,* he describes how it is possible for this sphere to become independent and despite this independence to be linked with the absolute. As an objective idealist he starts of course with the obvious answer: the true prime mover of every existing thing is the operation of the spirit. For all that, he is absolutely sure that this theorem cannot give a satisfactory answer since the absolute not only embodies and inspires the immediately perceptible appearances but also excludes them from itself. So the Whole cannot be pieced together from the sum of the particularities. In spite of all that, he tries to prove that these two poles are somehow still related to one another: For what is contradictory in an absolute and infinite manner (i.e. the Whole and the particular — M. A.) can be united only in the infinite. On the other hand, nothing can separate that which has been united in the infinite; what has thus immediately united will become contradictory in itself without further ado. In this respect unity and contradiction here mean one and the same thing. [Cf. Schelling: 'System der gesamten Philosophie', *Op. cit.,* p. 239.] What else is this if not the brilliant idea of the dialectic of immediacy, anticipating Hegel and pointing towards the structure of appearance and essence?

What we can also see here is that the struggle with the category of immediacy, besides having an influence over the phases of development of idealism, also gave impulses to the elaboration of the theory of contradictions. In a primitive and simple form there arise here the outlines of the Hegelian revolutionary thesis with respect to the 'identity of identity and non-identity'.

It is on this basis that Schelling gets to the point where he already suspects that illusions have a characteristic mode of existence. In one of his dialogues (dating from 1802) he writes as follows: Do you think it possible that what we call erroneous, inverted and imperfect might be real, or do you regard it merely as the product of the way we look at it? And his answer is: What I am talking about is not what the work is when contemplated in its individual forms in its separation from the whole. For indeed we obtain thus a completely inverted picture instead of a perfect work, false theses instead of true ones. [Cf. Schelling: 'Bruno, oder über das göttliche und natürliche Prinzip der Dinge', in: *Schellings Sämtliche Werke,* Stuttgart 1859, I. Abt., Bd. 4, pp. 121—122.] In other words, Schelling also notices that illusions have their origin in individual points of view and thus in their subjectivistic generalization. We cannot of course form an idea of the kind of connection that this presentational form (false because it is being formed only for the individual) has with the objective phenomenal sphere. But it is evident here, too, that his thinking is of a higher order than that of British philosophy. The sensualists simply want to eliminate illusions from the sphere of philosophical and social praxis, whereas Schelling wants to trace them back to their very process of origination.

We can properly appreciate the significance of Schelling's transitional role if we juxtapose it with Fichte's viewpoint. Fichte is still entrapped by the hysteria of his age. His subjective idealism is not only a German response to the British influence but also the result of an abstract petty bourgeois-revolutionary philosophy of history. For indeed the way he views the matter is that the bourgeois revolution has led to the liberation of the abstract individual and that is why immediacy is equal to the particular world of the abstract individual. Immediacy is thus a cognitive and active relation and it cannot even exist in an ontological sense.

The framework of this theory is the evaluation of bourgeois society where, according to Fichte, the unity of the essential ordering principle and the individual life has been lost. That is why the Ego remains within its own confines and that is why it cannot find objective bases. That immediacy is unreal and that it is only the intrinsic element of subjectivity is due to the fact that Unity has been lost in a historical sense as well. Fichte divides the development into four distinct epochs. The first of these is the unconscious rule of reason, the state of paradise in which the individual still enjoys no freedom because he is unable to effect an active realization of his subjectivity; the second is the period when the Ego awakens to self-consciousness and is confronted with the law of the community; the third one includes the breakup of the old community and the lack of a new unity — the state of guilt come true, or in other words, the age of negative freedom. (This would be the age of revolutions and the nascent bourgeois society.) It would be followed by the fourth state, described in its utopian colours, the world of self-justification come true and of salvation. In the bourgeois present, the third state of the world, immediacy holds sway. Only the individual as such is truly real, and this reality-creating function is the only tangible fact: Thus we come to an age which does indeed free itself from the instinct of reason but without being able to replace it with some other form of rationality, so that it is left with nothing real but with the life of the individual and with whatever pertains or relates to this individuality. [Cf. Fichte: *Grundzüge des gegenwärtigen Zeitalters,* Hamburg 1956, p. 27.] Fichte is here describing the petty bourgeois revolutionary illusions in the light of which the new order is not yet an objective mechanism with a logic that is independent of the individual. At this time he is still a child of the heroic illusions even if he has in some measure lost faith in his revolutionary ideals. He believes that individual freedom may be a principle capable of uniting the world. Consequently in his view the Ego, which posits the world, is the mere individual only on the surface: the individual is also immediately identical with the total movement, with the social whole. The particular individual is not separated from the mechanism of society but rather he becomes (mystically) identical with it. The individual and the great Whole are one and the same. Thus there is now no need for a sphere of immediacy in which individual lives and deeds may find themselves confronted with the movement of the Whole.

The importance of Schelling's decisive step seems to be more obvious in the light of this solution. It is with him that immediacy emerges as a contradictory sphere, as a medium forming between the Whole and the individual, confronting both the individual and the whole and linking the two.

It is only in the context of this conceptual struggle that the importance of Hegel's solution can be understood. Hegel is the first who indeed resolves the false alternatives of truth and untruth, of the whole and the part. We have already referred to some elements in his solution (immediacy as the form of existence of the mediations, the elucidation of the reflexive structure and of the ontological and epistemological content of this category etc.). But at this stage we only want to refer to the historical significance of the discovery. For indeed, in the *Phenomenology of Spirit,* he grasps this concept primarily as a historical category. In the course of the stabilizations of the self-creation of mankind the accomplishments obtained repeatedly become 'immediate' and are then transcended by the development. What he also realizes is that there occurs a form of immediacy in bourgeois society, separating the individual sphere of life from the whole, and that, in addition to projecting upon this sphere the configurations of total movement in an inverted and 'false' form, it can physically destroy individual existence so as to allow the total movement to make progress. "Consciousness, therefore, through its experience in which it should have found its truth, has rather become a mystery for itself; the consequences of its actions for it are not its actions themselves; what befalls it is, according to its being for itself, not the experience of that which is according to its being in itself; the transition (i.e. from the real into consciousness) is not the mere alteration of form of the same content and essence which we imagine now as the content and essence of consciousness, then as its own object, i.e. its intuited essence. Abstract necessity can be taken into account only as a negative, unperceived power of the generality, on which the individuality is smashed to pieces." [Verbatim translation. Lukács: *Der junge Hegel,* Zürich 1948, p. 609.] Consciousness thus is unable to take in the transposability of the configurations by which it is governed and with which it must work, although they function even if the individual is unaware of them. Thence it follows that Hegel also regards immediacy as representing the relative achievement of independence of the individual spheres of life, but here the relation to the total movement assumes a qualitatively more dialectical and more tragic form: "As a person in his individual work is already unconsciously doing general work, so when he does general work it is his own conscious object that is involved; the whole, as a Whole, becomes his own work for which he sacrifices himself, and it is indeed by this means that he regains himself from it." [*Ibid.,* p. 611.] According to Lukács, the genesis of these ideas is to be found in Ricardo's labour theory of value (private work and social work). Over and above that, what is essential for us is that the individual sphere of life is simultaneously the creator and the passive subject of the world it has brought about and which, because of the mediatedness of the many different kinds of movements which form the whole, it cannot survey in its entirety and which it still attempts to grasp 'as a whole'. Thus there arises under the influence of the real social totality an illusory, false totality which includes, however, the real one.

Methodologically, this is where the modern view of immediacy springs into existence, i.e. the dialectical outlook which already sees that the process of the origination and the way of the formation of what presents itself for the individual occur behind the back of the individuals simply because they occur in the objective process of the participation of all other individuals and, hence, they occur earlier even as regards time. The 'cunning of reason' is capable of explaining the form of immediacy, but the machinations of history are always cleverer. The ever new aspects for us (the 'formations of consciousness' inherent in scientific development and contentual self-cognition) can only be approximations. At the beginning of the *Phenomenology of Spirit* we find the following chain of thought: man experiences a certain relationship (object), and then his more recent attainments will correct the previous ones. It then turns out that what he has perceived was not the object itself but its being for us (its immediate manifestational form). Thus there takes place a dual process. On the one hand, cognition (experience) gets ever closer to the object itself and with this the 'being for us' gets increasingly nearer to things themselves; on the other hand, the things themselves also undergo some change and, as a result, they will again deviate from the chain of thought having already been formed. So mediation is, on the one hand, a subjective process (cognition) and, on the other hand, a historically changing, systematic arrangement of objective transposals, and truth can be nothing else but the conceptual apprehension of this historical process. The snag is, however, that this is precisely the process which hides truth. Only results are encountered. According to Hegel, consciousness is ruled by the necessities implied by the changing and developing mediatedness of things. Only necessity itself, i.e. the origination of the new object which presents itself to consciousness, without consciousness knowing what is happening to it, is what takes place for us, as it were, behind consciousness's back. With this, an element of being in itself is set in motion and this does not reveal itself to consciousness which is comprised in experience itself [cf. *Phenomenology of Spirit*]. The process of origination is hidden beneath the surface, and it can only be approached by consciousness in a 'phenomenal' and illusory way through virtual images.

Immediacy is simultaneously both an objective and a subjective category. It is not a concept of epistemology.

But for the time being this ontological category of immediacy is unable to explain the cause of the achievement of relative independence by the phenomenal sphere or of the increasingly fetishized superficial form in modern society. According to Hegel, immediacy is also the constant discontinuation of itself or the 'travelling cabin' of transmission in which there is no way to 'stop' or for superficial existence to develop.[16] This idea may be right as regards the immediate phenomena of nature, where every immediate relation is at the same time a substantive relation transcending itself. But in society, especially in everyday bourgeois life, there emerges a superficial layer which enjoys a certain amount of relative independence and does not discontinue itself. This objective superficial structure is elucidated by Marx. When he says of trade that this process 'is the result of productive capital, a process which arises from it and in equal measure sustains it', he is thinking of a superficial and immediate form of being which constantly discontinues but at the same time also reproduces itself in the very same form. Here immediacy is reproduced as a process, as a mobile category of being. Without this independent medium we could not grasp the special existential form of illusions as something which exists as an objective form of consciousness. It is only through the elucidation of this processual constancy that the ontological and epistemological contents of immediacy are separated from each other, which with Hegel were still organically (and hazily) intertwined. It is this method which was furnished by the Marxist praxis-concept of society as a whole. And with it the road is open towards a description of the ontological conception of reality and the objectifications of illusion.

We have seen Marx's explanation: Immediacy, as an ontological category, becomes clear through the discovery of the contradictory structure of praxis. Praxis, as an attempted solution, will appear again in 20th-century bourgeois thinking, but in a fetishized and simplified form, namely in pragmatism. Peirce and James in fact employ the description of the performance of everyday 'object-making' to form a category of epistemology. They identify the 'correctness' of the particular practice of the individual with the truth of the 'global man'. The basic experience is

[16] Immediacy is thus not some sort of a First, from where one may proceed and which gradually changes over into the negation of itself. Still less is it some existing substratum which gets started by means of reflection. Immediacy itself is this motion, writes Hegel at the beginning of the second volume of his *Larger Logic*.

furnished by the circumstance to which we have already referred several times, namely that one can act correctly even on the basis of false illusions. The immediate exposure of 'object-making' puts one in possession of a practical truth. Successful manipulation can be fixed also conceptually and used as a 'formula' which can be rendered independent. James lays a firm emphasis on this performance quality: "The truth of an idea is not a stagnant property inherent in it. Truth happens to an idea. It becomes true, is made true by events... Any idea that helps us to deal, whether practically or intellectually, with either the reality or its belongings... that fits, in fact, and adapts our life to reality's whole setting, will... be true of that reality." [W. James: *The Meaning of Truth,* New York 1909, pp. VI—VII.] However, this pragmatic view is based on immediate and individual praxis and from the outset excludes the criteria of total society, i.e., those which transcend immediacy. James consciously accepts the task of fetishizing the consequences of individual activity and, in opposition to Hegel's and Marx's theory of praxis which reconstructs a totality, asserts: "But whereas absolutism thinks that the said substance becomes full only in the form of totality, and is not its real self in any form but the all-form, the pluralistic view which I prefer to adopt is willing to believe that there may ultimately never be an all-form at all, that the substance of reality may never get totally collected, that some of it may remain outside of the largest combination of it ever made, and that a distributive form of reality, the each-form is logically as acceptable and empirically as probable as the all-form commonly acquiesced in as so obviously the self-evident thing." [W. James: *A Pluralistic Universe,* London 1909, New York 1928, p. 34.] The logical error is as follows: James denies the possibility of conceptual totalizability, whereas the Marxist conception postulated an objective ontological totalization which science will reconstruct subsequently but which would still exist in this objective quality of its totality. It also becomes clear from this chain of thought in what a roundabout way the neglect of the difference between immediate and mediate praxis will boomerang. After all, it was only through Lukács's aesthetic and ontological explications that the praxis-centred character of Marxism became once more a public property of philosophy only to be replaced later in the traditional and then dogmatic outlook by a Platonizing totality (i.e., by the abstractness of the historical process). The significance and influence of James's viewpoint — to return later in Husserl's 'international performance' — lie in his making the animated, many-coloured nature of immediate practice once again a problem of philosophy, in a false and fetishized form. He gave a distorted expression to a process in which man is not only the recipient but also the maker of truth. Kołakowski has an apt remark concerning this trend: "In this sense we can say that when we consider truth from a pragmatic point of view we are continuously creating and recreating the world: our cognitive relation to the world is its continuous 'making'." [L. Kołakowski: *The Alienation of Reason,* New York 1968, p. 164.] But it is indeed the social nature of the creative performance that does not show. It becomes clear at this point that this continuity cannot be formulated in its ontological structure not only because of the restrictedness of the class standpoint, though naturally the primary cause can be ascribed to that factor, but also because the continuity of the 'creation and recreation of the world' can be viewed only from the standpoint of the human race. Hence the nucleus of truth in pragmatism conveys only a fraction of the self-creative process of the 'human race'. But this fraction is extraordinarily important. This pragmatic outlook therefore becomes a recurring — and unsolved — question in modern philosophy. The massive flow of positivism and neopositivism proceeds from this same sphere of immediacy, even if its school-secrets are no longer stated with the same candour as at the beginning of pragmatism. The formulation of the performance and individuality of object-making does not point, however, in the direction of materialism. The value of ideas is not measured by reality (by the praxis of society), but by the functional role (acceptance or refusal) that they occupy in the sphere of private life; this role becomes the self-justification of false consciousness — but in an ideologically neutralized medium.

From the social necessity of immediacy and from the attempt at understanding it there arises a third solution: the intellectual system which tries to bridge over the poles between idealism and materialism.

CHAPTER 3

PRACTICAL CONSCIOUSNESS: SUBJECTIVE CONDITIONS FOR THE FORMATION OF ILLUSIONS

3.1. PRACTICAL CONSCIOUSNESS: SUBJECTIVE CONDITIONS FOR THE FORMATION OF ILLUSIONS

Let us reverse the course of our analysis. Until now we have paid attention mainly to the objective forms of manifestations of illusions, the domain of 'immediatization' which passes off on an ontological plane and ends in praxis. We made only a passing reference to the fact that it is after all the work of reflection which brings forth illusions from these objective forms of presentation. Now, however, we have to examine the process of cognition in which this formation of illusions, the false objectification of the presentational forms arises. Since these two processes are closely interlaced, this separation is of course also an abstraction just as the previous separation is. The basis of the problem is this: consciousness not only accepts certain relations among appearances (phenomena) and regards them as reality, e.g. the illusion of *post hoc, ergo propter hoc,* but it also enters into them itself with its activity. So practice cannot be separated from practical consciousness. We can say at best that they form the two extreme poles of the same process and that they also affect each other in this process.

The concept of practical consciousness is identical with the logic of everyday 'common sense'. It is the form of consciousness which controls orientation and activity and which György Lukács indicates in his *Aesthetics* as the category of everyday reflection. Since our purpose is not to analyze this particular form of consciousness, we must be content with tying down those threads in it which lead to illusory objectification. Simultaneously, of course, we propose to touch upon categories which Lukács discovered and some concepts which Ágnes Heller developed in her paper on the structure of everyday life. What is different is only the projection, our intention in analyzing these categories, since we will scrutinize them only insofar as they have something to contribute to virtual configurations. We will examine them as what Marx calls 'a transposed form of consciousness' which is the equivalent and the creator of 'inverted reality'.

3.2. TELEOLOGICAL THINKING AND THE REFLEXIVITY OF 'MAKING'

Everyday consciousness lives in the clutches of praxis. It is controlled, on the one hand, by thinking in terms of aims and means and, on the other hand, by the objective logic of actions and the current and spontaneous momentum of the things 'made', and the latter reacts on it with thought-creating force. Practical consciousness is the unity of these two positions, the conceptual form in a state of alternation and reciprocal transition. Let us look first at the effects of the teleological position.

Human action is controlled by a directedness towards some goal. But this means not only that things are completed earlier conceptually in our heads than in reality but also that we perceive and understand the objects themselves, too, only as functions of our purposes. Intuition itself is directed to such a goal. When we seek something, the medium which hides it becomes, as it were, invisible. Our mind is focussed on the picture of the object that we seek, and the only things that catch our eye are those which are identical with this mental picture. We may occasionally miss even what we are looking for, since, owing to the deficiency of our memory, our 'seeking picture', i.e., the form of the object in our mind's eye, is not identical with the object itself. (For example, when we fail to find a button though during our search we actually lay our hands on it; since, however, we misremember its form, we notice only that which is identical with this 'incorrect picture'.)

This directedness determines our emotional and conceptual position (the direction of our curiosity, our emotions and interests), and we see even the things themselves filtered through this subjective medium. But this also means that we read into objects of the present what is in the future, and that we treat what comes into our hands now as the potential vehicle of the later form. Things exist for us only to the extent that we can discover such a potentiality in them. The world exists for us not immediately but as the object of this teleological directedness, as 'potentialities' of success and failure, expectancy and victory, suffering and hope.

This teleological objectivity is, however, not identical with 'intentional objectivity', since the reality and the existence of things are not given by the directedness which only utilizes and accentuates them and grasps them as potential. In the course of work, i.e., the attainment of a goal, it becomes clear what belongs to the sphere of reality and what to the sphere of intentions. On the way to realizing one's goal one 'comes into collision' with the resistance and the otherness of the world, and with this already the illusion of the intentional 'creation of an object' is dispelled. One senses that one's teleological activity exists in an objective medium. The goals are filtered by reality which will let them through, modify them or even destroy them. And in this resistance we experience not only our subjectivity (such as the illusory nature of our plans, the objective value of our capacities) but also the objective 'weight' or quality of the world. So activity is a balancing process. Though things undergo changes in the course of realizing our goals, the results they yield are mostly different from what we expect them to be; they present themselves in their genuine aspects as contrasted with the abstract nature of our plans. To this Lukács adds that the 'otherness' of the result of our actions, the unknown component which alters the consequences, gains strength rather than declines with historical development. "As we get to know nature better, by the action of science and labour, this unknown medium becomes ever more blatant, and has the most important consequences for future human development. This unknown, uncontrolled area of social reproduction is not confined to primitive stages, but exists also at developed ones." [Th. Pinkus (ed.): *Conversations with Lukĺács,* Cambridge (Mass.) 1975, p. 19.] Science borrows from this recognized unknown, but neither is there any doubt that the as yet unrecognized unknown will be one of the components of illusion.

Teleological consciousness acquires at the same time a certain degree of independence. Contrary to the 'other kind' of answers of the objects it maintains its goals, chooses roundabout ways, resorts to the 'cunning of reason' etc. We could not otherwise realize our long-term goals. Not even the negative gestures of praxis can destroy certain fundamental or ideologically important goals, just as certain illusions will also endure despite the fact that reality refutes them day after day. We simply take no notice of this refutation. We even attempt to discover variations of our own illusions in the failures. One arm of the pincers of praxis is therefore teleological consciousness which alters objects by virtue of the fact that it grasps them 'differently' from the outset, as the potential of the forms which are registered in the goal.

The other arm of the pincers is the unrelenting reality of practice, the ruthless logic of the things 'made'. Workaday man is controlled and ruled over by more than merely his plans. In the greater proportion of the cases he must rely on the logic of things, he has to suspend his plans either temporarily or for good, and he must yield to the course of events. He may contribute at best to the current of processes which rolls along independently of him, while he himself also becomes the creator of this tendency which after all works against him. This passive 'making' has a peculiar 'reflexivity'. Man feels that whatever he is making is becoming his own. He will subsequently recognize the results coming of the things as an intended goal of his own. (This is so not only as regards rationalization after the event but also in cases of self-deception and self-persuasion.) In this process man produces his goals after the event so as to adapt them to a result of earlier origin. His collaborative work is a premising performance and so it inspires some goals too. Hence praxis-controlled consciousness works with a reverse teleology. It does not realize its previously elaborated goals but, on the contrary, its goals are created by the process of 'making'. Of course, in most cases this reversed order functions unconsciously. Self-deception, the system of spontaneous goals, is the false consciousness of compulsion under which man can regard as his own something that he is completing under the influence of external constraints.

That is why it is man's own actions which manipulate him most efficiently. A case in point is Hans Castorp, who on his arrival at the Swiss sanatorium immediately rebels against the environment which is alien to life. When, of necessity, he 'makes' the pattern of life of the others for a lengthy period, this collaborative *Mitmachen* slowly convinces him not only that this mode of life is nice but also that this is what he has in fact longed for all along and that this has been his secret goal. So man can be a party to the events not merely unconsciously but also through the self-manipulation of 'making'. (For that matter, dogmatism used this effect to re-educate people. Persons who asserted their counter-arguments were made to give effect to the very resolutions which they had criticized. Besides that this measure was necessitated by the political rule of the game — namely that internal arguments should not show outwardly — it also constituted a form of persuasion. The executants were forced by sheer moral self-defence to accept what originally they had done under compulsion.)

It seems that now the teleological plan, now the logic of 'making' acquires a form of control in practical consciousness. And this duality (the two arms of the pincer) has a lengthy past in philosophy. It is Kant who for the first time gives an explicit form to the power of 'made' things to influence consciousness. At the beginning of the *Critique of Practical Reason* he writes that we do not understand anything, except only those which we can also do ourselves. [Quoted by Gehlen: *Der Mensch, Op. cit.,* p. 318.] With Nietzsche this relationship is reduced, as it were, to a pragmatic formula: 'We understand only what we can make, if there is any understanding at all.' And finally in phenomenology, Merleau-Ponty identifies reality only with 'made' things: "There are no other things but the ones we make." [M. Merleau-Ponty: *Phénoménologie de la perception,* Paris 1945, p. 436.]

Apart from the ideological emphasis, there is a genuine problem implied in these comments, namely the autonomy of spontaneity. Everyday life does not control man as if he were living in a somnambulistic state and were carried along to the average, uniform way of life only by an unperceived current. Everyday life exerts its influence through the transposal of purposefulness. Man's original intention is something else than what the average tendencies have already carried into effect, and he consciously wants to turn aside from this course. But the results of his actions will still steer him towards that direction, and then the consciousness of his purpose will also accept it as its own by reason of the fact that he already senses that the new developments which were contrary to his intentions also correspond to some sort of a variant of the original purposefulness. The reason why this 'substitution' which is a refined and subjectivistic element of the formation of illusions can come into existence is because the cognitional form of things, the two alternatives of their 'being so for us' have an approximate coincidence in practice.

The known and recognized relationships and norms which we build into the system of our goals correspond in practice to the things which are altered through 'making' and thus become our own, even if *prima facie* they are in contradiction with one another. The two kinds of 'potentialities' become identical at the common denominator of 'reality for us', and if there still remains any difference between the variants it will be blurred by the emotive and conceptual factors (interests, prospective and emotional visualizations, prejudices, illusions etc.) of practical directedness. Our conflicts in life of course still arise mostly from the discrepancy that we cannot fit those results in our life which are not in line with our intentions and that in a concrete case the difference between these two presentational forms leads to our failure and our temporary crisis. So it is inevitable that we 'collide with ourselves' if the discrepancy between these two forms is too great. Everyday consciousness tries to avoid these conflicts and so it adapts the 'consciousness of purpose' to the answers of reality (to the 'result') whenever it is possible. After the event consciousness wants and understands this result of a different content which has come about as the dictate of reality and consequently not as the dictate of our intention as a 'goal of its own'.

This identification is an essential condition of our compliance with objectivity. There are also superior and more conscious forms of compliance in which we explain the causes of the different type of results on a theoretical plane and take into our service the link which has thus been disclosed. On the level of immediacy compliance is simpler because we adapt the picture of reality which is registered in our goals to the otherness of reality. Since teleological consciousness identifies these two kinds of directednesses, the two ways in which we are controlled are of the same category. This identification is true when we realize our goals. When we accept the 'other kind' of answers of life as our own we pretend that this is what we wanted to achieve all along. We are led by a false identification. But the element of identification remains common to both cases.

This substitution is most of the time not even the work of individual consciousness. We work mostly from the outset with concepts in which practical efficiency, and conceptual truth have already become identical in social use. They seem congruent, though ontologically there is one reference in the reflected picture and another in the element of reality which is used and made in practice. In the short term the difference between the known and recognized formula and reality which is made by aid of this formula becomes indifferent in biassed thinking. We may not even notice the characteristic of this difference which warns of the seriousness of the failure, because the essence of its structure is that socially the known picture of reality has been conflated with the objectively different results which it makes accessible. The range of possibilities of known and recognized relationships as well as of those which can be conceived as goals present themselves for us on a social plane, and we only individualize these possibilities and apply them to the requirements of our subject-being and change them thus into reality. The category of 'knowability' which Karel Kosik adopted has a teleological objectivity. The reason why we can make these conceptual relationships our goals is that they already present themselves for us as complete and implicit goals. We can make our individual requirements our goals and thus realize these requirements only in these concepts. Practical consciousness has no time for testing these known 'patents' also theoretically, and moreover, under the influence of habit, social conventions or the spontaneous demand for quick action, it would not even dream of doubting their truth. Since it became acquainted with them as relationships which, within a certain historical and social sphere, would lead on the whole to some result, and since these were already well-tried formulae also for others, it will not suspect that they are possibly false even when it comes to grief. It will feel that the flaw must have been in the individual way in which these formulae were applied, for indeed 'everyone' could make good use of them.

Thus the sphere of objectivity becomes wider for teleological practice. The 'objectless knowability' of the formulae, rules and norms which are registered in social consciousness also assumes this configuration besides the objects and the results which are in the process of being objectified. We might even say that the prevailing rules, norms and customs which are used in society are in some sense 'things in themselves' for individual consciousness which handles them as objective data, just as it is bound to accept the requirements of objects and processes also as objective data. This 'just as', however, calls for a more detailed explanation. First of all, these rules and norms are 'objectless', that is to say, they have no objective vehicles, so the imperative of objectivity presents itself only in a negative case, namely when we forget to observe them. (The various forms of greeting are conventions which have developed historically, but a failure to observe these conventions may bring in its train a grave conflict in some concrete case whereas the conscious breach of them can give rise to the explosion of social conflicts. Just remember the incident between Tell and Gessler in Schiller's *William Tell.*) As a religious or ethical norm, fasting also has such an objectless effect, and its compulsory character is ensured by society through the moral pressure of other people. However, we must also understand that in these 'objectless' norms and in the network of 'known and recognized' rules there were reflected active forms for the establishing of objective references, that these objective references then vanished in the course of time, and that the active forms became independent as rules and norms which have been stylized into something spiritual. (The sociology of religion has discovered that the underlying factor of the religious and ethical norm of fasting was the shortage of meat among the Jewish tribes, i.e. that originally there was an objective vehicle behind the spiritual norm.) If we emphasize the 'objectless objectivity' of the customary norms, we must attach special weight to our second stipulation. When we manipulate with the things and processes, their requirements are not only objective but they are also 'of the same value'. There is only one way in which we can achieve the correct result with them, and their utilization also lays only one kind of charge on us, a charge which we either learn or we cannot use them. There is a similar situation in the case of the norms, conventions of social origin, maxims, etc., but here a multivalent scale presents itself since there is here at least an alternative choice available to us. The 'objectivity' of the norm, however, remains objective here, too, for there will be the devil to pay if we fail to observe this norm. But 'observance' always has some sort of alternative character. I can do it wholeheartedly, formally or — acting freely within a certain range of possibilities — I can shape it in my own image, and I can even neglect it for a while. There is a wider scope, and what is more important is that the role of choice also arises here. All this is of course only a higher degree of objectivity. It does not diminish objectivity but merely complicates its effect. When the repudiation of the norms of feudalism occurred *en masse,* it did not simply lead to failure. First it conduced to the disintegration of the feudal institutions and then it resulted in revolution. And without the possibility of the alternative treatment and the altering objectivity of norms — social mobility and change, and indeed the objectivity, i.e., the historicity, of social conditions are inconceivable. So we must imply these two stipulations when we talk about objectless knowability and its objectivity.

But the process of becoming independent also means that it also stipulates the sphere within which individual action can be made a goal, but in such a way that the forms of knowability which have been offered and the true relationships of reality are not congruent with each other. Only a practical relationship functions between them. They are temporarily successful, but they function for a while even then as norms, as the objective instruction of known things, even when they already do not lead to success. Under such circumstances the reflexivity of 'making' adapts to the norms in social consciousness, and we cannot help trying to understand the different kinds of results through the traditional concepts and norms. Karel Kosik is thus right when he regards knowability as an obstacle to the truth of objective knowledge and self-knowledge: Stereotyped facts and items of knowledge do not allow one to go on and to seek thoughts in other directions; and moreover, one cannot even receive the otherwise conspicuous signals either. [Cf. Karel Kosik: *Die Dialektik des Konkreten.*] False totalization — as he calls this conceptual position — hides reality. The reason why knowability may prevent one from recognizing reality is that it could be for a long time the dominant motive of correct action. In everyday life the concepts of the correct and the true are confused since we hold that what leads to a result is — true. It will function as truth even when praxis itself has been further developed. When that time comes it will hinder social practice itself. The conflict between relationships grasped as goals and the results will be increasingly acute, up to the point when practical consciousness is forced to consider the feedbacks of 'making' in their independent nature, irrespective of the false explanations which present themselves.

The subjective possibility for illusions manifests itself in a historically always unavoidable 'gap' which is between these two different kinds of objectifications. The world of practical consciousness strives for homogenization, i.e. for an identification in which the differing contents of 'made' things and known and recognized rules and norms are congruent with each other. This congruency is guaranteed by practice since the result in reflection seems to vindicate the truth of the picture which is registered in the concepts and norms used. That this is only a possibility for illusion is because genuine and virtual relationships can alike be objectified in these conceptual forms, and they become illusions only in the comparison between the concepts used and their contents.

3.3. THE INVERTED WORLD: FINAL CAUSALITY

According to Marx, the most important feature of illusions is the reversal of the intrinsic conditions, i.e. the turning of reality upside down.[1] The objective background of this reversal is the transformation of the total movement on to the individual sphere of life, although this sort of projectional image is merely an abstract and objective form of presentation which consciousness turns into an inverted illusion, an objectified fact. We have already examined the first one, i.e., the presentational form, in our analysis of the sphere of immediacy, so now we are going to form an idea of how consciousness manoeuvres to effect this inversion.

The basis of this absurd view is the position adopted by practical consciousness: it subjectifies the 'commands' arriving from the objective world, but objectifies the answers to be given to them, i.e., it reverses the directions of objectification and subjectification. One realizes and objectifies one's inward subjective goals, but in practical consciousness this relation seems to be reversed. The inner motives seem to be outside commands arriving from an object, while the subjective positedness of this command is imperceptible. The command, 'This lock must be opened,'

[1] For example: All things manifest themselves absurdly in competition. The ready-made form of economic conditions as it appears on the surface in its own reality and which is consequently also reflected in the ideas through which the conveyers and characters of these conditions are trying to understand them is entirely different from, in fact the reverse and opposite of, what comprises their innate, essential however concealed base *(Kerngestalt)* and the relevant concept. [Cf. *Capital,* Vol. III.]

seems as if the lock itself demanded to be opened, and the operation does indeed verify this to a fault, because the search for the key, the manipulation with the picklock or the finding of the right tricks will indeed be dictated by the objective quality of the lock. But only because I want to get inside, this purpose of mine is projected to become the objective and material command of the lock; before I can get inside I must open the door. This sequence of immediate practice, namely that an earlier phase seems to be of primary importance for achieving the goal, is what creates the illusion that the obstacle becomes an exteriorized motive: subjective intention manifests itself as the command of an external object.[2] This outside command, however, is merely the 'relieving' illusion of everyday consciousness. It stems from the fact that, as a function of a directedness which is aimed at a further goal, a particular concrete action imperative will always 'pop in', which — in contradistinction to the deliberate and subjective purpose — will take on an objective shape. The consciousness of purpose had been objectified even before it was realized. Without this objectification no distant targets or long-range plans could be envisaged. But it already carries in itself the seeds of illusions.

But this illusion-creating objectification on the other hand is complemented by an illusion-creating subjectification. Things manifest themselves as subjective possibilities, as the sub-cases of 'what shall I do with them'. Fichte's and Schelling's 'creative intuition', the subjectivity of revolutionary practice, subsists also upon this fundamental relation of action ontology. The commands of objective situations become intentions of our own, giving the impression that their subjective directedness is the primary consideration. (This is the imperceptible adaptation to the 'logic of things' which we analyzed in the previous chapter, i.e., the attitude guided by 'making'. For example, no sooner does the careerist consciously take notice of the changed climate of the political situation than he already decides that his new boss is more likeable than the old one was and that this is why he wants to cooperate with him.) Now this interchange is obvious not only on the level of individual practical consciousness but it can also be seen on the social scale. Considering this from the point of view of social criticism, Adorno aptly writes that the concepts 'objective' and 'subjective' have completely changed places. We

[2] About this projection Gehlen writes as follows: The formation of motives in everyday life is generally transferred from the inner field of the tensions of wants, conflicts of motives and 'decisions' into the outside world. It is obvious that this process has a considerable relieving effect: our motivation relies on the indifference of the outside world from where it is controlled. [Cf. A. Gehlen: *Urmensch im Spätkultur,* Bonn 1956, p. 28.]

call objective what is not inconsistent with the phenomenon at issue, or more exactly what is identical with that expression of the phenomenon that has been unconditionally accepted, i.e. with the subjective. On the other hand, we call subjective what breaks through this façade and penetrates into the special experience of things, into the conventions of judgment — thus the objective. [Cf. Adorno: *Minima moralia*, Frankfurt/M.—Berlin 1951, p. 120.]

Why does practical and social consciousness reverse these poles? Because it is controlled by teleological consciousness, the practice that recognizes and realizes its ends in its means and the immediate conceptual form of this practice, i.e. the logic of final causality. The concept itself, the final cause, today looks like a mere curiosity of the history of philosophy. Voltaire, in his *Candide*, used absurd examples of this form of logical inference to ridicule the theodicy of scholasticism and Leibniz. Here the spuriousness of final causality can right away be seen through: according to Pangloss, for example, noses were made to wear spectacles, the sea was created to sail on. And what is also obvious is that Voltaire's absurd railleries represent merely the aftermath of the struggle which the exact sciences pursued against the dogma of the ontology of purpose. Yet, on the level of everyday life the logic of final causality has lingered on down to our days, without the possibility of detecting its absurdity. Nicolai Hartmann, in his *Teleologisches Denken,* called attention to the problem, though he, too, analyzed this method of reasoning as a historically outdated position, as a vestige of the anthropocentric view of the world, which persisted as long as man related each and every event in the cosmos to himself. Scientific disanthropomorphization replaced it with a reality-centred world picture which regarded even man himself as playing one of the subordinate roles in the total process and which no longer had the intention of conceiving our common failures as the revenge of Nature. But the man in the street still thinks along such lines: 'It is raining because I have not brought my umbrella...' The anthropocentric world picture lives on in our daily life like an atavism, since it is inevitable that the inherent relationship of the action between means and ends constantly reproduces the attitude of finality as the logic of practice. And this will give rise to the inversion of objective and subjective, inner and outer, and consequence and motive. Consider, for example, the phenomenon of opportunity. Regarded ontologically, this means an especially propitious coincidence of processes, a specific range of possibilities: a 'green light' for a particular politically or privately motivated action. Opportunities are suddenly arising possibilities that last for a very short period, never to return in the very same form. But their opportuneness is objective: they are not created by our wants or our expectations, despite the fact that it is 'for us' that they represent favourable possibilities and that without our intentions they could not even become 'opportune'. Yet they are real and existent possibilities; the barriers go up, and even in the event we should fail to notice the possibilities presenting themselves they would still be there — and we shall perhaps realize the missed 'opportunity' only at some later time. Opportunity is therefore just as much an objective situation as conflict is, but it is the form of our recognition of it, i.e. final causality that invests it with the form of relevance to us. To size up its objective manifestation and its demands we need a good eye, tactical aptitude, presence of mind and a sense of perspective. (This element in the self-revelation of reality is shown, for example, by the difference between the possibility of revolution and the way it is assessed. There are wasted opportunities for revolution, and there are cases where possibilities of revolution are not objectively present but only thought to be so, and these lead to failure. The failure also indicates that the point under discussion is an ontological phenomenon, for indeed opportunity is a compulsive objective configuration and failure to recognize it may cause one's perdition in the same way as the non-observance of certain counter-effects would.) Opportunity manifest itself by 'challenging' our teleological directedness. The manner in which we recognize an opportunity conceals this ontological structure and manifests itself to us with the logic of final causality. We say 'Opportunity makes the thief', as though the opportunity were there only for the purpose of turning honest men into thieves. Thus to make a man immoral things follow not their own logic but their relevance to the individual, and that is why they arrange themselves into an opportunity. Or to be more exact, goals which were originally of no interest to us would become real individual objectives, should the opportunity arise. Opportunity functions here already not with its ontological reality but as a form of consciousness. It is not a part of our own teleological goals, nor are we prepared for its manifestations. The purpose implied in it, i.e. its attainability, takes one by surprise: this is the exact moment when opportunity suggests the purpose. In this respect it affects us as a form of consciousness. It embodies an 'ought' *('Sollen')*. I must do something which I have had no intention of doing up to the present but which now when I am sucked in by the possibility of success suddenly becomes my goal. The pattern of thought behind the logic of final causality is as follows: the opportunity is here for the purpose of my doing 'this'. Things have come to develop in this way for my sake. They can tell me what my goals are because by some sort of divine providence their relevance 'for me' has aligned itself with me. (The watch has accidentally been left in the bathroom so that I can steal it.) In practical consciousness the opportunity is an end and a cause simultaneously. An end insofar as the thus far unknown and indifferent purpose reveals itself in it, and a cause if it encourages one to achieve this purpose and, as such, also becomes the creative force of the purpose itself. Thus the concurrence of certain conditions 'creates' truly favourable opportunities, but only in the everyday way of thinking does it become a final cause, because it reveals itself for me and so it relates to me and, in this respect, it offers itself as a means. It is indeed through the relation of means that practical consciousness reverses the poles of cause and goal, objective and subjective.

Hartmann conveys the meaning of this as follows: in our end-oriented activity first we envisage the conceptual goals and then, retrogressively, we search for the possible means of realization as causal relations which can be posited. Thus the reasoning runs backwards, in the opposite direction from the process of realization, from the end to the causes which can be used as means. For consciousness, therefore, the means are posited by the end. The end represents the 'cause'. Consequently the end and the cause are not only closely correlated with each other in the means but the end seems earlier than the cause. The end manifests itself as a cause. The logic of final causality 'reverses the relations of dependency of the process' (Hartmann) and functions in every means-oriented activity and not merely on exceptional occasions such as the unexpectedness of an opportunity. The means are determined from the angle of subjectivity and, in this respect, they even belong to it and form extensions of it, but as a result their objectivity vanishes — they become subjectivized; and *vice versa,* since the conceptually given goal figures as the index of the objective means, it will also be subjectivized itself; the goal — still in its conceptual form — will settle in the sphere of objectivity.

This is how subject and object change places in teleology.

Hence the secret of final causality as an illusion of thought is our relation to the means. Because our means are also double-faced themselves, they represent a *medius terminus* in which the subject's ideas become objective.[3] But in this respect, subjectivity is subordinated to the

[3] The means is the extrinsic middle term of the syllogism that is the realization of the end. Through it the rationality of the end manifests itself as such, so that it can maintain itself in this extrinsic other and by this extrinsicality. In this respect the means is superior to the finite ends of extrinsic expediency and the plough is worthy of more respect than are immediately the pleasures and ends obtained by it. [Cf. Hegel: *Science of Logic.*]

means: we are autonomous individuals merely in the abstract positedness of the goal, and in the course of implementation we are already dominated by the available means to which we must submit ourselves. And indeed, in immediate practice we see even the goal itself as a function of the means. Hence the reversal.[4]

What in the course of our enquiry so far has been merely an illusion of private action acquires social universality in capitalism. Here, for the first time, there develops an all-embracing social organism in which the decisive factor is its processuality. Every beginning is a result, every means at the same time conveys and offers an end, and the individual man, even if in alternative form, is guided with a goal factor which is drawn out into this means. We have already discussed 'true' reality being formed only after the event, as an effect of the overall conditions. Goals, too, will manifest themselves for the average man only when subsequently, in things taking shape between his hands, he can recognize his conditions and the goals which they offer him as possibilities. In capitalism the means are in control and suggest the goals. Tyrone, the protagonist of O'Neill's autobiographical and confessional drama, *Long Day's Journey into Night,* chances upon a role that earns him a great deal of money. After becoming rich he would like to play other roles as well, but to no avail, since the public demands to see him only in this familiar makeup. He is unable to find other roles because he has already hitched his wagon to the sole means bringing him success. And being the weak character he is, neither can he choose a different alternative. In the meantime his talent has gradually adjusted itself to this one style of acting till, finally, the public tires both of this role and of the 'one-dimensional' actor himself. Tyrone's actions and his goals have been determined by the 'means', i.e. by the chance opportunity of a role, and in vain does he want to create out of it a goal of his own and in vain does he search for his self-fulfillment. He becomes subject to the 'mechanism' of social laws and the possible goal given by them, which destroys him.

But subordination to the means is evident even in the simplest phenomena of work processes. There is, for example, the phenomenon of 'swing'. (Loggers and ballast packers exploit the work-easing role of the work rhythm as a help from 'outside', because they feel that the 'music'

[4] As Hegel put it: Of teleological activity we can say that in it end is beginning, consequence is antecedent, effect is cause, becoming is what has become *('das Werden ist das Gewordene'),* and that in it the already existent steps into the existence etc. [*Ibid.*, p. 347.] Lenin set great value on this remark in his *Philosophical Notebooks*.

of hammering tides them over their weariness. The swing encloses each series of actions, as it were, in a 'travel cabin': if we 'fall out' of it, not only our mood but also the successfulness of the work, i.e. its objective quality, will suffer.) And this once again is not an isolated experience but it also holds true of social processes. The spirit of a particular movement will stir even those who are unconcerned and carry them with it beyond the objective content of goals and interests as the objective effect of the spirit.

But the logic of final causality presents itself most significantly in the objectification of interests. Sociology regards interest-based outlook merely as an illusory self-deception, a prejudicial subjective way of seeing things and a distorting bias because it sees only a subjective category in it. [Cf. Simmel: *Soziologie: Untersuchungen über Formen der Vergesellschaftung,* Berlin 1905, p. 341.] Subsequently the so-called Frankfurt School, first of all Adorno [cf. his *Prismen, Kulturkritik und Gesellschaft* and *Negative Dialektik, Op. cit.,* particularly Part III 'Sein und Existenz'], Habermas ['Zwischen Philosophie und Wissenschaft: Marxismus als Kritik', in: *Theorie und Praxis,* Neuwied am Rhein 1967] and to a lesser degree Marcuse [*One-Dimensional Man* and *Reason and Revolution*] again raise the issue of the analysis of the category of value. Though they do not seek a way out in a subjectivizing direction, they still grasp the concept of interests as a part of ideology, as the practical projection of social consciousness and as a quasi-ideological sphere. These research studies have yielded important results concerning the harmony between interest-motivated spontaneous thinking and the social forms of consciousness, but they tend to lose sight of the fundamental discovery of Marxist philosophy that interests also have an objective characteristic just as the social conditions of existence do, of which they were created as emotive-mental forms. Anyone who acts consistently against his interests which have arisen from his conditions of existence will be ruined in the same way as if he were to wipe out his financial existence. Interests, however, have a different mode of existence. They become clear to us as objective possibilities and become active — genuine — interests at various degrees of their level of consciousness. For that reason alone it is odd that precisely the Marxist philosophy includes no account of an ontological analysis of this category. This lack is presumably due to the fact that dogmatism, similarly to bourgeois sociology, regarded it as a subjective development of intellectual quality. Dogmatism, aiming at the abstract necessity of social processes which are independent of people, considered that the analysis of this problem would be of third-rate importance.

Interests are also teleological in their structure, but this teleology already has a social transposal. Instead of immediately relating to the objects it is connected with them and the processes through a mutual relationship that people have with one another, or more exactly, through the intention of changing the other person's objective. When I go boating by myself I can designate my course and the island where I put into harbour as my goal, and I do not speak of it 'being in my interest' to reach this destination whereas if there are several of us sitting in the boat and, perhaps, each of the others chooses a different direction then I will certainly be motivated by an interest to talk the others into accepting the course I have chosen. Putting this in different terms, instead of being directed immediately to the object (simple teleology) the means here manifest themselves in the transposal of a social relation. My immediate purpose must be the persuasion of others, for the only way in which I can reach my objective goals is if I am successful in this. Accordingly, there are two teleological chains closely correlated with each other. The only way in which the objective means can be grasped is through a socially given causal sequence. György Lukács calls this relation secondary teleology in his ontological study. It is not things that I shape according to my goals, but other people's thoughts in order that they form the things so as to suit my goals. Accordingly, interests have various dimensions. We can speak of financial, moral, official or scientific interests, depending on the kind of social and human medium in which we work. On the other hand, Hegel is speaking to the point when he says that no matter what a man does, he cannot accomplish anything without interests. Our deeds can be shaped through the field of action of society, and the most abstract concept of interests characterizes both this sociality of the individual and also, motivated by this but also conflicting with it, his purposeful pursuit of his goal. So the difference is only in the content of interests, i.e. in the medium in which we measure our struggle for the use of material means against the other persons.

This dual teleology already assumes that it is not simply objects and people that face each other, but the means employed — i.e. material and intellectual media which can exert their influence only in society and which therefore in any case have a relation-creating and relation-conveying function — and individuals, who, even before they have concluded an alliance of interests, are already separated from or associated with one another by virtue of their conditions of existence.

This is then what marks the difference between the simple teleology and the social (dual) teleology. Robinson Crusoe, with his primitive tools, could harness only Nature to serve his ends, so his tools were of the 'bipolar' type. A man moves one end of a simple lever and the other end lifts the heap of stones blocking his way. One pole is turned towards the individual and the other one towards Nature. The means which social man uses are, however, at least tripolar: only through the other person (through his resistance, his defeat or his help) can they be used individually. So it is not as if it rested with the arbitrary decision of the individual to what ends and how he would bring influence to bear on the other person as regards the use of the means. The bounds of the second teleological condition are given from the outset (within the scope of possibilities). The conditions of existence and the relations of ownership determine in advance the direction in which the owner and the one who is excluded from ownership (both using the same means) may look for their respective goals. (The landowner, the tenant and the peasant work with the land as a means, but the land has different ways of mediating their mutual relations. The landowner handles it as the source of rent, the tenant as a profit-making opportunity and the peasant as the object of his livelihood in the course of his daily work.) So, on the one hand, the means mediate people's social relations, and on the other hand the given relations determine the limits of the functions in which people can use their means. And the relation of interests can function only within this sphere: it is not the creator of these conditions but only their irradiation. The kind of immediate goals which the peasant has in view are put into a certain frame by this given relationship he was born into. Within this frame he may come to alternative decisions and, depending on his individual facilities, his cunning and his partners, manoeuvre so far as this range of possibilities permits, but the limit of his possibilities is necessarily given. That is the source of the objectivity of interests. These possibilities are objective in the same way that there is a certain objective hierarchy of success among the alternatives presenting themselves. One solution is more advantageous, the other is less so. Yet this objectively outlined action imperative is jointly determined by two factors: the universality of the conditions of existence on the one hand, and the acting person's individual situation, his suitability, his particular desires and hopes on the other. Individual interests represent that objectively manifesting course of action related to other persons following which one can, from one's individual position, attain the success which is socially realizable.

Individual interests exert their influence through the transposal of recognition and individual decision. There is one objectively optimal solution in every situation, but it depends on the acting individuals whether they can recognize their alternatives or pursue imaginary interests and, if they do recognize them, whether they can or in fact want to choose precisely that particular optimal solution. And besides, interests pertain to the future and, therefore, an element of uncertainty also has to be taken into consideration. This element may manifest itself objectively in the available solutions proving to be successful only to a certain degree of probability or subjectively in the individual being unable to hit exactly upon the optimal solution. (In the case of choosing a career one must decide now about finding a normal livelihood in ten or fifteen years' time and an occupation to suit the person one will then be.) But the stochastic probability is affected not only by our present inclinations but also by the trend of the subsequent economic development and any change which in the meantime may occur as regards our individual requirements. (Occupations which are regarded today as successful ones may become backwaters in ten years from now, and it may also happen that the person who at the beginning found his self-realization in the job of an electrotechnician will in a later phase of the development of his personality look on this occupation as slavery.) In interests, therefore, the objectively existing form of action which is given as a possibility is closely correlated with its recognized variant which has been made into an individual goal and which is a more or less exact or inexact approximation of that objective possibility.

The reason why recognition, consciousness of interests, is only approximate is that it is mediated by two objective 'filters', namely by the objective hierarchy of interests and the individual interpretation of this hierarchy. A great many intercalated goals and interests mediate the maintenance of our existence, and the question of what it is that one finds important or prefers in this sequence of mediations will depend on his individual decisions, personal character and attitude. Whether someone accepts abject poverty to the end of continuing one's studies and by doing so attaches more importance to his long-range and intellectual interests, or takes a well-paying job instead, is a matter of his individual scale of values. Recognized interests, however, do not yet constitute realized ones because also in this case the pressure of the conditions of existence prevails. The selection of values is not simply a question of character. If a person, contrary to his original intentions, is forced to discontinue his studies because he has to support his ailing mother then the decision has been interfered with by the objective constraints of the essential conditions of life. (Though it is conceivable even in such a case that, in opposition to this constraint, he takes a contrary stand on the issue.) The man who acts 'against his interests' will clash with the average accepted interests by invariably choosing higher or lower interests.

In everyday life, however, the scale of interests is simplified since we are for the most part dominated by immediate, momentary interests. These filter through our fundamental interests of which most often we may be conscious only to the extent that everyday interests let them through. Here is one source of false consciousness. We may indeed have long-range, prospective objectives, but our long-range interests which involve making decisions far in advance never materialize in their pure forms. They present themselves only in daily transposals — hence in false forms of recognition. Only on rare occasions, in a crisis or a revolution, does the true structure of the fundamental interests of individuals and classes become clear to them. (That is why, for example, class-consciousness, i.e. the process of one's becoming conscious of the fundamental interests and conditions of existence, must be inculcated in the consciousness of the working class and in fact of all classes from without.)

Hence individual interests are realized in the recognition and the decision, whereas class interest exists in two forms. On the one hand, it is an average direction of action developing from decisions made by the members of the class, consequently a result, a tendency which others can also recognize and follow: an objectification of interests; on the other hand, it is an alternative possibility arising from the outset from their shared conditions of existence, that is to say an objective starting-point. These two constituents are generated by one another. Class interest is not an abstract necessity floating above the heads of individual people but it is the objective configuration they have formed of their possibilities of action, a configuration which is therefore always a larger unit than the 'chances' of an individual. So there is a certain random relation between the optimal steps which follow objectively from the conditions of existence and the reflection of these steps in the interests. Imaginary interests may, on occasion, prevail for long periods, though these errors will later be corrected by objective interests. Due to this correction, the average decisions will be balanced eventually. An even greater discrepancy is found between the conscious apprehension of interests and their objectivity. It is not necessary to recognize our interests in their pure form since their realization is possible from false motives and genuine ones alike. The teleological element therefore cannot be eliminated from the objectivity of interests. Interests function only through this teleology and exert their influence only in the transposal of recognition and decision. But this is only a functional recognition. It is a 'cover-category', as it may be both a genuine recognition and an illusory one. Max Weber describes how the missionary character of the Islamic religion was changed by military interests. In the course of the conquests there emerged a need to retain a non-Muslim and therefore inferior slave people who could be exploited. This fundamental interest, however, manifested itself in the reshaping of the tenets of Islamic religion, and in its direct form it was not even known about. [Cf. M. Weber: *Economy and Society,* Vols 1—2, transl. E. Fischoff, H. Gerth, *et al.,* Berkeley—Los Angeles—London 1978.]

Here the category of final causality emerges again. The motives of everyday life are usually formed not so much by conscious and at the same time genuine interests but rather by the adoption or the rejection of the appeal of certain methods, steps and possibilities as regards value. Thus, for example, a particular social connection becomes first attractive, then useful and therefore capable of being turned to good account, and it will turn out only subsequently that we were in fact pursuing our interests. Büchner, in *Danton's Death,* splendidly describes this objectifying mechanism by changing the resonance of Danton's name. The revolution swings to the left, but the masses do not yet consciously sense this change in the situation. They suddenly believe the earlier gossip which has presented Danton as a pleasure-seeking, avaricious egoist. The 'moral devaluation' of Danton, which would have been inconceivable to them somewhat earlier, though they knew the facts, is now carried out by the masses themselves. They objectify the development of their interests and situation in their leader's change in value. This is the psychological explanation of why narrow-minded people are capable of using devious tactics and a fantastic skill to the end of asserting their interests. Their actions are not directed by rational and circumspect consciousness of interests but rather by the spontaneously emerging value relations of situations, things and possibilities, and they have a highly developed flair for surmising the change of these values which are indicated for them. (In a love affair any woman can with the greatest ease entangle her partner whose intellect is superior to hers; the able merchant sometimes cannot give an exact account of why he buys some goods and why he wants to get rid of some others, he is simply reacting to the changes in the apparent value of things.) Hence in interests we focus on the mediating possibilities or people and, usually after the event, discover our goals in them, as though these goals had dictated our directedness.

The illusion of final causality, therefore, presents itself, even before we have set ourselves our target, in the form of attracting us, wanting us to work with it and to move in a certain direction which likewise presents itself in a 'peremptory' manner. It seems as if things existed to serve us or to determine the direction of our action. The world reaches us through the filter of interests, and its original relationships are reshaped by the structure of secondary teleology. This is the way in which the subjective and teleological category of interest is transformed. The particular attractions and value changes of things invest themselves with the directedness of our intentions.

The manner in which conscious apprehension of interests occurs in established social practice is of course not so simple. On the one hand, in our stereotyped processes of action we do not even need to be steadily conscious of our interests since the institutions in advance take over the role of conveying them. On the other hand, in the case of conflicts of interest we are forced to awaken ourselves to the consciousness of whether it is worth getting involved in some unpleasant clash, or whether it is 'worth our while' to avoid it. So in everyday life our interests present themselves not only in a spontaneous way (in our taking notice of value changes) but also in the hierarchy of the consciousness of interests. In our regular daily life this hierarchy is limited since there is hardly any difference as far as our spontaneous and conscious interests are concerned. It is only through the media of which we have not even any knowledge that we can comprehend our fundamental interests. In the case of crises or revolutions, however, we are forced to awaken to the consciousness of these interests and even of those joint interests which are connected with other people. Under these circumstances the objectivity of interests comes true also socially, as the visible movement of the classes.

The changes in the value of the means employed, i.e. whether things or the steps taken manifest themselves appealingly (in a desirable form) or disagreeably (in a form to be avoided), are projected on to things or possibilities by our social position, or to be more exact by the objectively given source of our teleological directedness. But we can hardly notice these sources. Things which appear for the 'position' seem first as vehicles of value and will work in their functional characteristic only afterwards. This interchange is brought about by the element of final causality inherent in the logic of interests. In the activity of classes this objectification of values may become independent. Its basis and its point of departure has an ontological character. The given possibilities are either suitable or unsuitable for a class to realize itself. [Cf. *Conversations with Lukács, Op. cit.,* p. 30.] 'Suitability' is an ontological fact which is explored and accumulated by the successfulness arising from the average of innumerable individual decisions. But this ontological suitability will acquire that form of consciousness of the objectification of values in which there is no need to get to know the whys and the wherefores. Worth may even in itself become the object of the consciousness of values. Late feudal society precluded the bourgeoisie from the possibility of achieving political success. Sectarian ideology and then Calvinism created an independent value out of this objectively formed separatedness. They discovered in it the circumstance that, as matters stood then, bourgeois economic development must go its own separate way and that it must work according to a more Spartan and more rigorous ethic than the 'nonchalance' of the nobility. First came the objective fact of the separatedness, then its practical value, and finally this value acquired both conceptually and ideologically a quality of independence in Calvinist ideology. An element whose content is unknown but which proves to have a mediating function and the property of setting goals not only becomes valuable but may also become an independent intellectual scale of values, an ideological system. It is here that we shall have to look for the order of successive steps in the development of false consciousness and ideology. We have only to realize the triple gradation of this process: first the manifestation of the objective media (conditions of existence), then their reflection in values, and finally the rationalizing, arranging and rendering independent of these values to form an ideology which directs praxis. And since these three steps are an abstraction from genuine social praxis, they actually abstract from it, so the achievement of independence by the values and their arrangement into a theory or ideology will fix not only the relations of action of a particular group but will also contribute to the self-development of mankind as a value. After it has been formulated as a scale of values it becomes an instrument in the self-creative process of mankind, irrespective of whether the given era can or cannot use it. Several hundred years may have to pass before people can return to it once more. The possibility of their normative role was, however, originated with their becoming a scale of values.

Yet, through the transposals of the three gradations, things and possibilities manifest themselves not in their original objectivity but in their functional 'being for us'. Seeing and thinking in terms of interests alter things. First of all, because of final causality it reverses the relation between cause and effect, starting-point and result, or to be more exact, it brings the connection which has a reverse manifestation in praxis to our consciousness and registers it as such. Secondly, original objectivity suffers a change in the form of the category of values; man will even add a few subjective elements to the disagreeable or agreeable things. He regards the useful thing not only as nicer but also as contentually more perfect. Interests, as objectifications of thought, help to register illusions. They are the originators of conceptual forms in which we adopt these false forms of presentation and project them on to reality. Besides illusions some genuine relationships are also registered in these distorted forms — in a false configuration. Expressions which attach value to interests, when they go beyond the limit of the merely particular, are the guiding motives of genuine social movement in which their objectivity is apprehended. But the forms of *a posteriori* rationalization mostly consciously sanction certain particular class-related conditions, though in a false conceptual and ideological form. We will call this structure a simple expression of interests and, as rationalization of interests, we must separate it from both the spontaneous consciousness of interests and the higher forms of ideologies.

Yet interests are not only subjective 'direction indicators' but also guiding motives for the objective 'making' of social existence. A distinction must therefore be made between their ontological function (their social role and their objectivity) and their epistemological content (or the role they take in its reflection). Ontologically the fundamental interests of the classes take as their starting-point material reality and ascribe their practice to this. Success is the criterion of their correctness, independently of the extent to which as conscious interest they correctly reflect the mediatory routes. From this point of view it makes thus no difference to what extent the forms of consciousness of interests (scales of values, ideologies) provide reality with a new character and to what extent they confirm or deny its illusory forms of presentation. Historically, however, there is even here an important difference between the illusion-destroying tendency which is inherent in the interests of the progressive classes and the illusion-maintaining function of the conservative groups. (At the end of the 19th century the striking worker had no idea of the sort of illusion that could serve as ground for wages to become the basis for exploitation. He simply did not accept this illusion, and his refusal was based on empirical denial and not on an epistemological, i.e. contentual and scientific investigation.) But we ought not to treat this matter mechanically, since we know of conservative interests, too, which have been conducive to ideological achievements with progressive final results. (Remember the role that the bourgeois historiography of the Restoration period played in the development of historicism.) At the same time, this conjuncture takes effect through a great many transposals. The agrarian communist sects of the Middle Ages opposed the Church on account of the tithe, and the ideological manifestation of this conflict of interests was mediated by the vestiges of village communities, the ethical outlook upon life of the formal 'give so that I may give'. The progressive nature of their ideology did not stem simply from the role of the 'progressive class' but also from a whole series of traditions which had become in an objective sense formal and illusory by then. Ontological indifference towards illusions therefore means only that it depends on the historical phases of development whether the interest of a group or a class help to create illusions or to destroy them — and by doing so to build new ones. From our point of view this is then the decisive aspect of how we are to form an opinion on interests as such. Their mode of existence controls material praxis.

Epistemologically the situation is entirely different. Here we measure the objectivity which manifests itself in interests against real things in its contentual quality. The medium which exists in the teleology of interests and the real medium can blend with one another and become an illusion, while consciousness of interests conceals all this because it objectifies it. Illusion is measured by the reality of success. Thus we go beyond epistemological relations and at the same time remain within them because we maintain our illusion that in them we reflect the objectivity of things. Original objectivity is changed then by interest orientation which invests it with a new quality of value. That is why — from the point of view of interests — objects, people or methods which serve our interests (or are judged to be useful) become more attractive, better, nicer and worthier of imitation, whereas those which harm our interests will be disagreeable and bad. This requalification of values is the essential character of teleological illusions. Behind every bias, prejudice, partiality, selfishness and self-deception there is this illusory form, a role in which interests take effect unconsciously and establish new values. We discover this false increase in value (or loss of it) only when our interests change together with our situation and we can at last see the object in its 'original' light. But its illusory forms are present in nearly all our actions which are motivated by interests.

Hence the model of the conceptual reversal of reality is final causality, or more particularly the way in which the structure of the consciousness of interests works. However, this relation of consciousness is of mediating nature in any case. It is not an independent form of consciousness, nor is it merely a practical relation. Now it is the activity itself and now its emotional and conceptual control. Ideology and false consciousness represent one of the media through which interests assume a conscious and objectified form, though interests do not get entirely dissolved in this sphere, nor does ideology acquire a quality which is identical with consciousness of interests. Interests have become conceptualized but it is this medium in which they become an objectification that can reach independence.

3.4. THE FORMALISM OF EVERYDAY CONSCIOUSNESS

Illusions acquire final objectification either in the conceptual formation or in the process in which certain stereotyped forms become independent. We do not notice the essential difference behind things which are formally identical. We accept illusions which have been produced by analogy or we ourselves produce the analogy which performs the identification. And likewise, we do not notice the essential identity behind the formal differences. In general, as we have seen in an earlier context, for everyday consciousness the regularity of formedness (for example its symmetry, rhythm or its fitness for expression in a closed form) or its applicability as a formula is identical with the substantiality of the given appearance although ontologically the form, the formedness is indifferent towards the categories of essential and unessential, of law and chance. This can best be illustrated in artistic forming. For example, besides the fact that the forming of rhymes takes effect through the euphony of the line-ending accord, it also imparts to the emotional and conceptual content which is expressed in the lines a certain universal character because the cadence of the accord gives the impression of regularity. But both real as well as false emotions and thoughts can produce this impression, and only the entirety of the poem will determine whether the recipient receives or refuses it. The rhyme alone and in the way it is formed always intimates such a more universal and more substantial content. Proverbs and those conventional norms of current usage which have been turned into rhymed tags have a similar structure. The way in which the expression is formed represents the essentiality of the content, the fact that the given relationship or norm cannot even exist in any other way.

The formalism of everyday consciousness is adjusted to the analogue.[5] That is all it notices: it sees only that with which it has already familiarized itself in the similar and already known 'other'. It is controlled by formal similarity and it operates with the validity of this similarity, remaining indifferent towards the contentual difference. That is why it deems that that which is formally similar is more real than that which is similar in terms of essence. We have already seen how this intuitional adjustment is complemented by the thesis of formedness and substance. Practical consciousness believes it is discovering the essentiality of things in their formedness and in the peculiar resemblance of this form. Coincidences repeat themselves two or three times; for example, in the case when the pupil is questioned and caught off his guard on successive Tuesdays, and he sees in this a system or a formedness, suspecting at the same time a latent law of nature. Most types of illusion are born of this formal relationship and rigidity in the common consciousness as 'everyday forms of thought' (Marx). The *'post hoc, ergo propter hoc'* is the result of this outlook in the same way as analogical thinking is also controlled by it, or as manipulation also deceives the individual through the production of similar formal relationships by creating an intuitional uniformity in which the individual and particular characteristics of things and with it their different-in-kind essentiality also disappear.[6]

But this formalism is more than something which leads only to hollow illusions. It is also an indispensable expedient of everyday life. Looking up contentual items, if we were to do it in every single case, would be too

[5] At the end of the thirties an actors' contest was organized in England to decide who could best imitate Chaplin. The great comedian also participated in the contest but he was placed only third behind those who gave a better imitation of him and who were consequently more like him than he himself was. Günther Anders describes a tragicomical case as follows: Hollywood discovers an actress with a fresh complexion and they predict a great career for her. But before she is allowed to stand before the camera she is persuaded to have a plastic surgery in order to acquire a face which is completely like the prevailing type of stars. In other words, she has to get rid of the quality through which she could become unique and hence different, which is in fact the condition on which her success depends. With the surgery performed her face becomes quite similar to the prevalent face type, but at the same time she becomes uninteresting and so the producers drop her. [Cf. G. Anders: *Die Antiquiertheit des Menschen,* Zürich 1960.]

[6] A. V. Gorbatov describes in his memoir of *Years off My Life* [London—New York 1965] how on the occasion of a certain May Day the veteran communist in the village wanted to remove the shabby red flag from the top of the council house to replace it with a new one. Since, however, the outlook of the authorities and the peasants had been determined by the manipulated ideological cliché of the 'enemy is tearing down the flag', his gesture was seen as if he were going to tear down the flag. The veteran was of course arrested amidst general indignation and outrage.

long a process and would hamstring quick decisions, whereas the orientation which is based on formal marks shortens this action series. That is why social practice develops by and large effective forms of action which function on the basis of probability and serve as controls of quick action. And as regards their statistical average these effective and also conceptually evident relationships, precisely on account of their probability, probably register mainly real or at least usable connections. But the trouble is that through this use they become formalized so much that they will hide their outwardly similar but individually different essentiality. For indeed, intuitional orientation in the effective rules only concerns itself with the matter of whether things which present themselves or which are being handled fit in or not with the given form of thought, and makes their truth and the quality of their substance contingent on this integration. So the evidence is the conceptual stereotyping of a real relationship, its formal generalization on a probability basis, a generalization which praxis verifies in the majority of cases. Hence the formalism of thought does not stop at the abstract manoeuvre of consciousness. It can be true because on the whole everyday life vindicates its correctness. But what it vindicates is only the formal capacity, the validity, i.e. tautology. The careerist always knows which gesture or measure he has to offer as a valid currency, but those who accept this at face value pay heed only to the usability and validity, that is to say they concentrate only on the formal capacity. Correctness and validity constitute a social currency which vindicates and renders itself usable only in its quality as regards form.

This formalism, however, has still deeper roots. It presents itself already in the working process, when with the tools we concentrate on a single and formally apprehensible property of objects instead of their individual characteristics. The tool itself also uniformizes to a certain extent and, in this respect, performs a formal abstraction. (In the mill the different sizes of the grain are not taken into account. The machine is set for an 'average' [formal] size.) Standardized production, an essential condition of production in modern capitalism, further increases the process of formalization, i.e. formal outlook. The 'same', the analogue, precedes the individual, and the valid form comes before the content which fits into it.[7] No sooner does

[7] According to Günther Anders the real simply becomes a reproduction of the pictures which it produces, because it is an ontological law of economics that there is only the standardized product and that the individual piece does not exist. And he writes later: The real which originally served as model for the pictures that were made of it has now been altered according to the external appearance dictated by the picture. Reproduction forms reality. [Cf. G. Anders: *Die Antiquiertheit des Menschen,* Zürich 1960, pp. 179 and 190.]

the mask of a particular famous star appear on the screen (in a sense as the double of some existing type) than thousands of spectators with a similar hairdo and make-up are watching and enjoying this picture in the theatre. It is the rule of the universal equivalent which creates this formalization in social intercourse. The commodities and with them forms of action must be brought in line first with the valid relation of exchangeability. Their individual characteristics may be taken into consideration only to the extent that the constituents of this form of equivalent, if they stand apart, 'disappear' or else become 'invalid'.

Validity is the stereotyping of a real state of affairs. But its functioning is subject to the condition that it becomes a conceptual cliché of the outlook and that it determines the way in which people can see reality and thereby their actions. And on this level it is valid also as an unconscious motive. (Gordon W. Allport made the observation that university students who fought against racial discrimination, become themselves — unconsciously — captives of these prejudices in another connection. Socially formed stereotypes penetrate deep into the individual mind and permeate even a conscious attitude which is opposed to them.) [Cf. G. W. Allport: *The Nature of Prejudice,* New York 1958, p. 309.]

So formalization is, on the one hand, a social tendency and, on the other hand, simultaneously the necessity and the limitation of everyday action. It has a tremendous relieving power and, at the same time, prevents us from getting near to the reality of things. The taste for originality deteriorates in it. A copy always bears a closer resemblance to the original because it can be used better, because it has already been tried out. For indeed, the original with its independence and its unrepeatability, that is to say with its lack of analogues, is not only incomprehensible but also useless. It cannot be applied again in the same way. We can use it only when it already resembles itself. This is the contradiction under which everyday practice and scientific cognition labour. They are urged on by the necessity to elaborate stereotypes, whilst progress is conditional on breaking through clichés and discovering both the special differences of things and the substance which manifests itself in them. This is the contradiction which lies behind the endeavour to formalize the world and the centuries old attempts at reducing it to mathematical terms. From Leibniz to the achievements of the modern philosophy of language, mathematical logic or structuralism, we find this endeavour in various forms. We find it as a real and usable relationship and also as an illusory overgeneralization. Modern cybernetics would be impossible without such a formalized picture of reality and, at the same time, the illusion of reducing things to mathematical terms is almost automatically generalized from this successful application since it seems that it can be applied to all walks of life.

No doubt, in the age of manipulation the increasing formalization of average thinking and of the scientific outlook also contributes to the objectification of illusions. So we must admit that Marcuse is right when he believes that in the fashion for the methods of operationalism and linguistics he has discovered the methodology of hiding the essential relationships. [Cf. H. Marcuse: *One-Dimensional Man, Op. cit.*, pp. 150 and 180.] We must, however, point out that every formalized structure of thought can express certain ontological relationships of reality, and that formalism is not identical with the simple distortion of individuality. In connection with the theory of Wittgenstein, but with a more universal claim, György Márkus writes as follows: "But even the abstract logical propositions (e.g. 'pvp') convey a certain ontological content. Whatever propositions we substitute for the propositional variables, these schemata will give true results because they exhaust all the alternatives of the possibilities of truth and correspond to all alternative possibilities within a determinate distribution of possibilities." [Cf. his Introduction to the Hungarian edition of Wittgenstein's *Tractatus Logico-Philosophicus,* Budapest 1963, p. 60.] Formalization can take place on different levels and each of them expresses and summarizes different abstract possibilities or real — ontological — relationships. (The structuralistic ethnology of Lévi-Strauss steps 'nearer' to the formalized relationships of the elements of reality, whereas mathematical logic sums up the furthermost layers of these possibilities.)

The formalized method grasps these interactions, which interlace one another by means of inherent relations, in their abstract quality. But the abstraction of these interactions is a relationship which exists within a certain sphere. The philosophical problem arises at the point when the question is where the limit within which we can use this abstraction as a rational one lies. At present we can give only a negative answer since the more complicated the totalities which combine with one another are, the greater the ontological obstacle to a successful formalization will be. Namely, this inherential relation gives rise not only to the interplay of different qualities but also to some new qualitative characteristics, creating thereby a new configuration of 'being so' as well. And this new quality can be grasped through formalization only in its details and its fragments because this method, just as qualification, can total the relationships only in the course of a particular aspect. (Or it can put the different qualities before us in a form reduced to a single aspect.) The ontological difficulty which prevents the successful applicability of this method beyond a certain level of involution is formed by the complexity of the totalities. A work of art, for example, is a complex totality in which there is a condensation of the spiritual energies of a particular age, the subjectivity of the artist, the manifoldness of the material, revitalizability, the various energy levels of the effect and a number of other determinatenesses. The reason why this complex totality which has some significance beyond itself cannot be described in its fullness by the categories of information, structure or formal relations is that the manifoldness of the representation would be lost and reduced into a single aspect in the course of formalization. Thus, for example, it is undeniable that numerous more general aesthetic characteristics of a particular artist or trend could also be brought to light by analyzing the languages of form. But at the same time the unsophisticatedness of this analysis which is implied from its nature precludes the possibility of describing the characteristics of the language of form, a characteristic which revives and changes its appearance here and thus. The uniqueness of Zola's *Germinal* and the effect it had on world literature cannot be deduced from the general or quantifiable characteristics of naturalism, although it has numerous features in common with the formal characteristics of Zola's earlier and later works. But the individuality of the breakthrough becomes clear only for an analysis which can give careful attention to the human contents of this work, that is to say in the light of a complex analysis, and only then does the analysis of the linguistic form become a rational abstraction. According to our hypothesis, the culture of the primitive tribes in Africa or South America is also a complex phenomenon which we can grasp through the description of the structure of rites, artistic relics and tribal customs only in a single aspect. The special historical significance of this culture, the characteristic which could be described in an African mode of production on the analogy of, shall we say, the 'Asiatic mode of production', is necessarily left out of this aspect. And it is due to this complexity that progress has slowed down to the extent that this process seems to the observer to be almost outside time. The formalization of complex totalities today itself requires a complex apparatus, but not even this complexity can pass the ontological limit where formalization is no longer possible. As a specialized method of science, this formalism yields results in a case when the aim is to measure and project the different totalities against and upon one another. (Thus, for example, the increase or the decline in the number of telephone calls may be the index of a particular political situation. But this situation cannot be grasped in its complexity even if we set up innumerable similar formal indices.) None of these indices can simultaneously show the relevance of its content to and the persistence of this relevance in all the others. Modern bourgeois philosophy at the same time tries to find further possibilities in this direction. There are emerging more and more formalizing methods of investigation, and even universal models of thinking and developing branches of science that become absolutized. (For example, Lucien Sebag divulges the workshop secret of structuralism when he sees in the analyses of linguistic philosophy the model to be generalized as the modern world picture. He sees not the, in many respects new, precondition of an analysis of a specialized branch of science but a substitute for a comprehensive picture of the world. This tendency towards absolutization is already what the age requires.)

We cannot of course simply set mathematical logic on equal footing with everyday consciousness, although the two methods have many common features. The difference is to be found in the quality of manifestness besides the level of abstraction. Practical consciousness thinks in terms of the formulae of action and creates a practical manifestness with the formalism of these formulae. A concrete relationship can be either included in or excluded from the stereotypes, proverbs and practical formulae. Manifestness functions as an intuitional filter, even if this form of intuition is created by ideological concepts which are reduced to manipulative formulae through everydayness. (It is the dictate of fashion and its uniformizing effect which serve here as a model. The sign of normality and indeed of phenomenal nature is the observance of certain laws of form. The violation of these laws degrades the individual to an 'impossible creature' in the opinion of the average man. And this is so not only with respect to clothes. Even the styles of the high culture of thirty years ago have become public property by now. The concept of the 'inferiority complex' may be the subject of a porter's conversation, and it may even mean an attribute which endows him with some rank or qualification. The process in which the concept acquires an everyday form has at the same time turned into a filter of intuition, an evaluative indicator which shows a false essence.) The formal methods of philosophy are, however, a great deal more abstract. They are instruments which function not by intuitive but by logical systems or complex analyses. Manifestness relates here not to the practical success of immediate intuition in terms of action but to the successfulness of a logical process of thought. 'Integration' means here the fulfillment of certain logical conditions. The point under discussion here is not simply that this logic functions with abstract forms which are devoid of content, for indeed the content is also abstract itself, but that it is the truth or falseness of the possibility of the relationships of reality which is the only contentual and at the same time form-creating element. So the accusation of formalism cannot strike home, for indeed the system starts from the principle that it examines the relation of abstract contents. In our opinion the criticism is appropriate only with respect to the scope of application: where do we find the kind of relationships in which these formalized and abstract contents can absorb and express the true connections? Or in different terms: where is the manifestness of the mere 'it fits in' or 'it does not fit in' sufficient? At the same time I think that in this abstract sphere the relation of practical consciousness to reality is still the basic methodological index. Success does not indeed function on the basis of an immediately practical and usable standpoint, but as for example the abstract possibility of the 'if this then that' formula (implication) it still scores this kind of variation of success. And moreover, logical truth depends on the — conceptually — successful or unsuccessful fulfillment of these conditions. The model for the superior configurations of formalization is the method of practical consciousness. Like everyday practice it approaches phenomena with a view to success. But here the success is valid within the autonomous system of the logical or formal system.[8]

Accordingly, what in everyday life leads to the production of illusions in the spontaneously formal creation of connections of consciousness uses a subtle system to establish conditions on a higher level for the inexcludability of illusions. On the basis of the formal picture of manifestness, illusory social relations can figure as real ones in the same way as true relations. The formalism of consciousness is fetishized with this absolutization.

[8] Adorno, with sarcastic malice but in my opinion appositely, points out that in Husserl's theory of manifestness the everyday practice of the bourgeoisie is stylized even in the immanence of the phenomenological method: The primacy of the method corresponds to the primacy of the social organization. The way in which known things are capable of being handled becomes a criterion of itself through logical classification. That which does not fit into this classification can present itself only as a 'datum' on the edge of life and must wait for something to 'fit into'. If no such thing arrives, it is thrown aside. [Cf. Th. W. Adorno: *Metakritik,* Frankfurt/M. 1956, p. 150.]

3.5. IDEOLOGY AS QUASI-REALITY

To what extent does ideology belong to the sphere of illusion? The problem was simple only for dogmatism. Ideology and illusion are concepts which correspond to each other, the former being the theoretical formulation of the latter. Stalin's theory of ideology thereby in fact adopted the simplifying traditions of the Enlightenment. Stalin's view, similarly to the 18th-century critiques of religion seeing only clerical fraud in ideologies, held that ideologies constituted a lie on a social scale. He derived their structure immediately from the interests of the classes and regarded them as forms of prophylactic or after-the-event rationalizations which had been created with the intent to deceive, as contrasted with Marx's conception in which ideological reflection was always closely connected with the objective logic of the social situation. With this the theory of ideology became subjectified. It lost its objective unit of measurement and the effective forces of its origin and development. It became an epistemological category, or even more so, a psychological concept, namely the instrument and phenomenon of deception and self-deceit, i.e. the creation of intentions and illusions. But this subjectified understanding made contact not only with the Enlightenment's conception, but also with the view of modern philosophy. Sociology — with its bias towards social criticism — has, ever since the phenomenon of manipulation presented itself and hence not without reason, regarded ideology as the theoretical form of illusion, not only as a mere expression of interests but also as the conceptual (theoretical) reproduction or indeed production of social illusions. The subjectivistic element dominates here, too. Illusions are produced by thought, whereas ideological configurations have no ontological, only epistemological meaning.[9] There is no doubt that mani-

[9] Although Karl Mannheim's theory of ideology represented this position as early as the end of the twenties, this interpretation became a universal trend of sociology only round the beginning of the fifties. Adorno noted then that "ideology, this socially necessary illusion, is today identical with real society". [Th. W. Adorno: *Prismen,* Frankfurt/M. 1963, p. 21.] Habermas terms ideological consciousness "the immediately made carbon copy of false reality". [J. Habermas: *Theorie und Praxis, Op. cit.,* p. 314.] Herbert Marcuse, on the other hand, derives man's encirclement by illusions from the objective logic of technological progress, since it is due to the technological revolutions and the results of comfort that he himself creates the false world and ideology. [Cf. H. Marcuse: *One-Dimensional Man, Op. cit.,* p. 190.] The opposite of this absolutizing theory is Raymond Aron's theory of 'freedom from ideology' which inferred from these same premisses as well as from the decline of the philosophical systems and the passing of 'totalitarian methods' the extinction of ideologies and with it the impossibility of producing illusions. [R. Aron: *L'opium des intellectuelles,* Paris 1958.] The common starting-point of the two extreme views is the immediate identification of ideology with illusion.

pulation psychologized against traditional and relatively objective illusions ever more aggressively and elaborated an apparatus which immediately shaped everyday consciousness. The ontological structure of this apparatus was indeed eclipsed by the subjectivistic effects of consciousness, and it is therefore indisputable that this critical attitude had a certain legitimacy. But the essential character of either the historically developed great forms of ideology or its present configurations cannot be explained merely by this manipulative intention. The practicism of dogmatism was placed here on a common theoretical platform with bourgeois philosophy. But on account of this we did not get an answer to the most important question. Even those relations of existence assume a conceptual form in ideologies which have not been intended by anybody and whose real content is not known either to those who invented them or to those who utilize them. This form exceeds the individual's intentions and the scope of his knowledge: it is the conceptual objectification of his conditions of existence.

Thirdly, since Stalin's conception fetishized the concept of the 'basis', it was unable to grasp the ideological sphere ontologically. It made the historical processes and also the movement of the classes independent of the individual persons' activity and endowed them with independent movements, a circumstance which led to the personification of history, to the tacit revival of Hegel's conception of *'anima mundi'*. Behind the façade of materialism there lay concealed an idealistic conception of history. According to this interpretation, individual events always become sub-cases of the autonomous movement of the historically necessary *'anima mundi'*, and it was not evident that at the same time individuals were the prime movers and the creators of this process. There thus developed the paradoxical situation that the very theory which denied the independence of the spiritual spheres tacitly hypothesized the spiritual independence of history. Behind this duality there lay a practical necessity. It described the intellectual medium of 'gross manipulation' (Lukács), according to which the purpose of ideas is to adapt people to the 'casting moulds' of society.[10] But the revolutionary idea that people sometimes make these casting moulds before the relevant conceptual

[10] The contradiction between involuntary Hegelianism and the conscious opposition to Hegel in Stalin's philosophical conception has already been described by many; among others, by Lukács, Kołakowski and Kosik. Most recently a substantial study by Iring Fetscher has analyzed the causes of this paradoxical phenomenon and the views of those who discovered this contradiction. [Cf. I. Fetscher: *Karl Marx und der Marxismus,* München 1967, p. 95.]

frameworks was strangely enough excluded by precisely this revolutionary theory.

Dogmatism, the Enlightenment and modern philosophy thus concur in the absolute separation of reality and idea and consequently also in the hypothesis that ideology is illusory. It is not our purpose to outline a comprehensive theory of ideology, so we must be content with examining the intellectual forms of ideology only in a single aspect, namely the extent to which they can be grasped in their ontological structure and where within this basic form of their existence do the possibilities for creating illusions manifest themselves. The primary function of ideologies is to stimulate the objective process of social praxis. They are integral parts of this process and therefore we can describe their characteristic features only ontologically. At the same time their epistemological quality, a conception which shows reality through the filter of a characteristic outlook, still conveys the possibility of creating illusions and, under concrete circumstances, their truth. So we have to examine the connection between these two dimensions.

Marx's reform, as contrasted with the conception of the Enlightenment, was to resolve the absolutized separation of the real and practical spheres and the intellectual one. Human practice is the creator of both the materialistic relations and the spiritual configurations. Marx considered ideology a part of practice and could imagine its separation and independence only in the unity of human praxis. In the fourth Feuerbach-thesis he formulates this idea as follows: "Feuerbach starts out from the fact of religious self-estrangement, of the duplication of the world into a religious world and a secular one. His work consists in resolving the religious world into its secular basis. But that the secular basis lifts off from itself and establishes itself as an independent realm in the clouds can only be explained by the inner strife and intrinsic contradictoriness of this secular basis." [K. Marx: 'Theses on Feuerbach', in: K. Marx and F. Engels: *Collected Works, Op. cit.,* Vol. 5, p. 4.] So Marx acknowledges Feuerbach's merit in understanding ideal bodies of thought as a projection of man's true relations. But Marx believes that this discovery is only a starting-point, since it only describes the process without revealing its causes or its origin. Why does the ideal world separate from the material one? And Marx indicates aphoristically that the answer is to be found in the unity of the reality of human practice and in the 'discord' of this unity. Thus it is not as a subsequent restoration of a different, spiritual world that ideology is in communication with men. It is not a simple rationalization. Its life and product are material and they are created by the reality of praxis.

What does the discord of the secular basis mean with respect to the mode of existence of ideology? It means the contradiction which we have already touched upon in several aspects, namely the relative difference and the subsequently developing unity of the immediate reality of individual practice and the equally real structure of the processes of society as a whole or of a class. Only as a rare exception can the individual person identify himself immediately and mainly through his purposeful actions with the trend of either his class or the processes of society as a whole. But his particular life is tied to the existing generality not only by the fact that he builds and is then controlled by this totality but also because he is 'reconnected' with it through the transposal and the 'different kind' of medium of ideologies. The historical function of ideologies is that, at least within the bound of possibility, they elevate the individual beyond his mere particularity and that, by means of the different forms of false consciousness, they attach him to the general tendencies of development of the group, class or of the human race. This duplication is distinctly manifest in the notions of gods in tribal development. There were as many gods as tribes, but these otherworldly substantialities laid down the requirements of the common existence of the given group. They fixed the norms which were not the same as the goals of individual people and which could be safeguarded only in this way, in this abstract form. With the expansion of communal existence 'local gods' gave way to religious notions of a more universal validity which were free from local restrictions. Their more universal validity and their normative influence are ensured by their breaking away from the starting-point and therefore by communal existence assuming a more abstract form. (Gehlen sees clearly that the world-creating and therefore, by implication, universally valid gods were also there before, but there is simply no room for them in the life of the primitive communities. They will become dominant when there is a greater difference between individual and communal existence and when the tribal organization breaks up [cf. his *Urmensch im Spätkultur, Op. cit.,* pp. 254—255].) The basis of the general relationships and processes which are (falsely) reflected in ideologies is 'generality' which acquires relative independence also in reality, even when the depiction of this generality can never be adequate and when it always takes place in a mediate conceptual form. (That is why Marx uses the epithet 'ideological' to denote conceptual distortion.) This can mean also the apparency of the ideological view of things. What people read into reality through the 'spectacles' of this is a nonexistent, virtual relationship. But when this false form is transformed into motives for actions a transformed picture of true 'generality' may, as we are about to see, appear in it.

At the same time, ideological forms create only the possibility of contact with 'species-being' or class-relatedness. Moreover, in a given case these forms may even separate us from them. The function of declining ideologies or of those which are becoming reactionary is precisely to maintain particularity and to obscure man's 'species-being'. The core of reality in false generality has decreased here to a minimum. And it is also clear that different ideologies can effect this connection with different qualities. In the arts this energy which goes beyond particularity takes place in the artistically created illusory sphere, in the separate world of the work of art. In legal and political concepts it occurs, whether in traditional or prospective forms, with the help of laws which can be put to use in economy. But the individual's life would be inconceivable without such a high degree of control. He cannot live in a merely particular way, even if consciousness stays within the narrow contours of his selfish life. His socialness is the elemental condition of mere survival. Therefore our life is interlaced with ideological elements. Men's conscious and teleological activities are constantly pervaded by these elements which have have been popularized and simplified to routine patterns and without which we could not formulate our motives or even perform our gestures. Whether a husband forgives his wife for her misconduct or feels that only a divorce can make amends for the injury, the alternative poles of his decision, whichever he chooses, are provided by various ideological sources. The solution that he chooses means at the same time the selection of a particular ethical standard even when this step is immediately dictated only by custom, social etiquette or prestige, and the individual is thus not even conscious of his decision being of the nature of a norm. There are different ethical, religious and ideological presuppositions behind every pedagogical slap administered by parents, and their gesture can be 'made' only from these ideological intentions. (Whether the child gets slapped on account of his having beaten up one of his classmates or because he has allowed himself to be beaten up implies two different pictures of the world, though people have not the remotest idea that they act in this way.) These ideological motives also affect economic decisions, although it seems as if purely material interests were the relevant decisive factors. Whether someone in the 18th century puts his inheritance in the bank and draws an annuity on it or opens a small firm is seemingly contingent on where the better opportunity presents itself in the given situation. The decision is in fact influenced by numerous ideological elements, and it is only through these elements that people become conscious of their financial interests. The hoarding of money, for example, is a dishonourable thing for the puritanical Calvinist, and the question is consequently decided from the outset by the religious precept. But even if these ideological elements could not serve as such models of actions (since there are still no developed alternatives of the thought) the decision would still be partly ideological. In the boom period of industrial capital each individual person was confronted with the alternatives of whether to choose the new but risky way of living or to stay with the old one. And his decision was not influenced by the abstract tendency of the development of the classes but it was made as a trend by the individual decisions themselves, and the choices were prepared by a new conceptual order of motives which moved the individuals to change their mind to this end.

Ideologies which are built into everyday life as its constituents are parts of real praxis, although as motives they do not accurately reflect the movement of reality, and furthermore it is possible that in a contentual sense they become the right motives precisely through their illusoriness. The fact that the Puritans refrained from the accumulation of money was a false motive in the sense of its immediate significance. According to their special theory of salvation only active business could constitute the test of election in the hereafter. The way of life of an annuitant rated as a passive and consequently 'sinful' way of making a living, and from this illusion which was false as regards content there followed the correct movement which could redirect social practice from the obsolete forms of monetary funds and credit to the more dynamic industrial capital. Or to be more precise, new practice and individual alternative were totalled, to become this theoretical motive.

Practice is not only the consumer but also the producer of completed theories. Any particular theory is created not when it acquires a philosophical formulation of a high order but when in everyday decisions difficulties present themselves which can be recognized and surmounted only through a new conceptual position. Only on the basis of a new ideological approach can new phenomena be recognized at all. In the old ones they either remain invisible or form only a disturbing factor which has to be eliminated. The Church of the Middle Ages — supposedly on the basis of the incorrect reading of the *Vulgate* — carried on a fight against all forms of money interest, seeing the basis of exchange in the principle of the 'just price'. Transactions were allowed to be concluded only at production costs and therefore without a profit. This ideology was of course not only a mere error. It fixed the norms of limited exchange — and not of trade — which were pursued in the sphere of village communities between neighbourhoods. But at the height of the Middle Ages this principle already encumbered business life and so the Church was obliged to grant more and more concessions to this social pressure. These roundabout ways, however, did not solve the problem. A sense of social deprivation developed with the desire not to circumvent this rule but to become master of it. This was achieved by Calvinism which not only sanctioned the ethical nature of interest and profit but which even saw the proof of divine election in it. This new view, however, made its way only gradually. It emerges from the sectarian movements of the Middle Ages, polished through the many transposals of heresies, because increasingly more people found themselves in the illegitimate situation of making illegal profit. And at the same time there was an increased demand on the part of society that this illegal and already inevitable situation should be not only legalized but also recognized as a stimulating motive. So it transpired that not only the individual decisions were totalized in which more and more merchants took the risk of making a profit but also the motives themselves. Collective action selected the most expedient principles in this totalization. Thus a new outlook had sprung from totalization and later a new religious ethic which went on once again to affect people's decisions from without. Even more obvious is the relationship between the everyday practice of individual people and the development of ideologies with respect to the emergence of the principle of land rent, for instance. The inventor of the differential land rent, James Anderson, was a practising tenant and raised what he experienced in his own praxis to a theoretical level. But this invention became an ideology only with Malthus and Ricardo. The 'pure' scientific invention which had sprung from practice was impracticable. It could become a useful theory, an ideology, only by being tested and polished against the demands of class praxis. This is the accumulation of ideas. Social polishing at the same time leads this theory from individual practice to class praxis. In this accumulation it is encrusted with more and more specific elements, and it breaks away more and more from unique and individual experience. As a theory it can become independent.

In the course of this accumulation, ideology becomes the authentic reality of a particular age. It becomes a quasi-reality since people take for reality the guiding concepts and norms which live in their ideas and views. The forms of presentation are determined by the conceptual framework which gives the clue to the understanding and the interpretation of these forms and thereby to their mode of existence. And this framework which intuition has raised to the level of reality becomes more genuine than the original substance of life. For the French revolutionaries the Greek and Roman heroes represented not merely a costume. They really believed that their world was assuming the forms of ancient democracy, and for a while the success of their actions even seemed to bear out this illusion. But this 'error' of thought and intuition is not a simple illusion, for indeed it proceeds not from the objective forms of presentation but from the conceptual explanation which has been 'read into' and projected on to it. On the contrary, its correctness is guaranteed by practice. So this intuitional illusion is illusory only with respect to its husk. With its help the struggle of the final rise of the bourgeoisie can be conducted with a historical correctness. In a more general form this means that ideology as something which controls the self-realization of human integrations offers a true picture in its result and an untrue picture in its immediate conceptual and intuitional capacity which reshapes reality. It is a quasi-reality; partly because there is no other reality which the individual can perceive and interpret or there is at best a different picture of reality which has been reinterpreted likewise ideologically, and partly because practice in the long run does not contradict itself if it accepts this reality which intuition offers. Though the individual person or a practice which is set on a short term can time and again bring to light the falseness of these forms of presentation, historical truth will nevertheless assert itself through its characteristic feature which misleads individuals.

And secondly, ideologies mean not only the reinterpretation of the already existing conceptual system to serve a political interest. Their quasi-reality arises not from the fact that the scientific and normative categories which have already been developed suddenly represent a different function of value with which they lend a new colour to reality. Kołakowski is right in seeing this new orientation of values as one of the characteristics of ideology. [Cf. Leszek Kołakowski: *Der Mensch ohne Alternative,* München 1964, p. 26 ff.] Thus, for example, the same theological principles serve as weapons for both the Lutheran and the Münzer parties, but within these conflicting factions they represent different orientations of values, and their instrumental roles also become different. What still outweighs this role is that important ideologies further the discovery of a new context of reality and give a conceptual framework for the expression and perception of this context. (With the scientific world picture, a new theoretical attitude also arises which is in striking contrast to the conception of reality based on medieval transcendency.) But of course the new set of concepts and with it the new picture of reality which has once again been accepted as authentic does not mean at the same time that the false element has been destroyed root and branch. The progress of ideologies does not proceed in a straight line towards the scientific picture of the world which is devoid of anthropomorphism. (Robespierre's subjectivism, for example, is 'more untrue' than the rationalism of the Enlightenment.) The new conceptual and intuitional form of reality is 'more correct' only with respect to praxis since it directs the average of different integrations more successfully. This development in its final tendency also proceeds towards the approximately correct conscious perception of the total action of society.

The new form of illusion is thus also included in ideologies as an element. (That is why Marx sometimes refers to ideologies as distorted reflections, as pictures of reality which have been turned upside down.) But this apparency is only a form, only a conscious husk of the unconscious 'making' of mankind or of class development. This is the conscious praxis of 'they do not know it but they do it'. And that of the knowledge of 'I do not know'. The unknown and unknowable element stems from the circumstance that ideologies have two sources, namely real tendencies accumulating from individual actions and the conceptual and intuitional forms of a common directedness which gives rise to these tendencies. Ideologies thus formulate more than what can be understood from individual praxis. The unintended common elements, that is to say, the directedness of integrations of different levels (class, nation, human race), the command of their conditions of existence, their prospective desire and their rationalization are also included in the known and knowable concepts and norms of ideologies. (That is why the reality of principles which finds expression in philosophical propositions and in works of art is more than what those who formulate these principles know, want or understand. The effect of the ideology which functions, i.e. has become practice, is always 'different' from its abstract intellectual content.) But this 'surplus' is not perceptible for those who consume this ideology. This, too, has provided matter for the dogmatic notion that ideology is the simple rationalization of class interests, that its validity and its truth will vanish together with the classes, and that its application is only a deception. But if that is so, then how can we account for that which persists in every ideological form, the norms which can be used in subsequent ages and the works which provide people with artistic experiences?

No doubt, the immediate source and ontological role of ideologies is to control the practice of groups and classes. What is essential in this respect is their role of holding together and controlling integrations. But ideologies, through their connections with the groups, also reflect the 'directedness' of the whole of society and of the development of mankind, a directedness which is — unconsciously — included in the immediate orientation of values. So when on the basis of György Lukács's remark, in which he refers to Marx, we define the ontological locus of ideologies as embodying the realization of the opportunities of classes for struggle and of their desires and employable means, we must also make it clear that this immediate class-relatedness can be conceived only in relation to the total development of mankind. Important ideologies are conceptual forms in which the directednesses of the individual, the class and the human race are formulated and are in transition to and pervade one another. It is these conceptual forms, says Marx, in which people become conscious of their social conflicts and fight them out to the finish. But this fighting to the finish would not be possible if its directedness stopped short at the immediate interests. The self-realization of the classes is also part of the development of mankind.

The form of 'fighting it out to the finish' also means that it may be both a means employed by the progressive classes in their endeavours and the conceptual and coercive apparatus of the institutions of the ruling or declining order to be used for fostering the compliance and curbing the rebellions of the majority. (The Church and the religion of the Middle Ages as well as modern manipulation belong to the latter group, whereas mysticism and then the Reformation as well as the 'religion of reason' to the former.) It also follows from this widely known fact that ideological forms must be such wide generalities and ideal conceptual models as can encompass both their conceptual autonomy and the given tactical tasks (of a different nature). This idealized characteristic is true of all forms of ideology, but their final polarization was created only by the bourgeois revolution, by the ideal paroxysm of the *citoyen* concepts. It was in connection with the French Revolution that Marx pointed out the necessary and obvious relationship between bourgeois reality and *citoyen* abstraction, namely the paradoxical fact that it was in the name of an abstract and ideal concept that the limited bourgeois interests were won. The function of this ideality was to make use of its formal and abstract character in ensuring an easier success of individual and group interests. The creation of the formal code of laws suggests, for example, that recourse to equality before the law and the opportunity to live according to a system of laws are open to all and sundry. In actual practice, however, it gave the green light only for the bourgeoisie, with a view to financial success. The binding nature of the ideological forms were felt only by those who were subject to these forms, whereas those in higher quarters attempted to circumvent them and simultaneously render them valid for everyone else. The generality of ideologies implies both their apparency and at the same time their functional reality, that is, the practical stimulating role of the existing or the coming order.

The key issue of the quasi-reality and quasi-falseness of the ideological forms is overgeneralization. No false ideological form exists which would not include some factual context of historical reality. But even the truest ideology expresses this core of reality in the transposal of overgeneralization. Emphasis pertains to the very essence of ideology. What is an important element of reality in its 'noumenal' connotation is formulated as the only truth in the medium of ideology and in the logic of a class's 'fighting it out to the finish' and it even assumes a false generality in its weight, content and conceptual form. But this overgeneralization is an ontological necessity of historical practice. Such overgeneralized and therefore inspiring ideas are needed to carry on a successful social struggle since only important motives can spur people on towards great feats. That is why revolutionary ideals always go well beyond formulating the essential tasks and the 'true' relationships. Without heroic illusions the Jacobins could not have accomplished the 'limited content of their struggle' (Marx), and without this ideological 'surplus' the classes could not have been taken to the barricades. The motives which have an inspiring effect on people can be expressed only in absolute terms.

In ideologies, however, this overgeneralization presents itself not in its true form, as a necessity of practice, but as an expansivity of the idea. The source of this autonomy is the logic of the self-determination of the classes. The only way in which the groups can form and conceptually understand themselves is when they are related to a greater unit of themselves. The reason for this ontological relation is not only that the ruling or the revolutionary classes want to set the other groups behind themselves but also because it is through the structure of their existence, i.e. of their social affinities, that they are connected with this greater unit. The self-determination of the classes and the ideological expression of this self-determination (overgeneralization) also imply the self-development of the human race.[11] But this 'surplus', i.e. this 'greater unit', is

[11] In the *Grundrisse* Marx uses an economic example to elucidate this connection with the greater unit: Capital can relate to itself only if it posits itself as a greater unit, for example as part of world capital. This duality (i.e. national capital and world capital) in which capital can relate to itself as a being that is alien to itself is a damnably real thing in this case.

both ideological since it can be grasped only in the forms, semi-recognitions and mystifications of false consciousness and real because it is a projection of an ontological relationship, i.e. of the structure of the total movement of society, a structure which encompasses and moves the individual, the group and the human race in different ways. This is where we find ideological overgeneralization and this is where we have to look for the contradiction of the illusory, however ontologically still true, objectification of this overgeneralization.

The formula is of course not so simple as regards the events of history. Certain overgeneralizations become corrected in almost no time when they are confronted with reality. When they turn out to have been false, they will be replaced by a more adequate form of idea. (Consider, for example, the metamorphosis of Christian ethic, when the persecuted religion became a state religion. The rigour of private ethic continued to exist for a while, but its indefensibility soon came to light and there arose the duality of the principle 'render unto Caesar...' in the separation of legality and morality, thereby 'correcting' the overgeneralization of ideology.) There are, however, massive ideological forms which persist even when practice contradicts their formulae for a prolonged period. Notwithstanding that the inherent contradictions in the ideas of the French Revolution had already come to light in the peasant revolt in Brittany, these ideas lasted on the whole till the 1830 uprising, albeit with a paler intellectual enthusiasm. Ideological elements can become even more alienated if the class structure grows rigid, when it becomes a caste system. The doctrine of the transmigration of souls in Hinduism is for example, as it were, a 'closed' configuration of thought, a configuration which is simply immune to all worldly denials since it is solely the individual himself who makes his future destiny, and consequently there is no possibility of importing the 'refutations' of reality into this doctrine.

Overgeneralization is thus a characteristic mixture of the genuine and the subjectivistic elements of thought which are objectified by social practice. But whether it is the genuine and therefore true content which will come to the fore in this mixed configuration or the false conceptual material which serves as a framework is a function of historical praxis at any given time. It is possible that the class cannot achieve any result with the laws that are true in an epistemological sense because they do not possess communal meaning, because they do not have the inspiring and intentional content which is necessary for the logic of the fight to the finish. (We can see this in connection with the origination of the theory of land rent. The differential land rent, as we have seen, cannot be used in its pure form but only in its 'ideologized' form. It could find its way into the practice of society as Malthus's theorem, since it was through this practice that there emerged the theory which 'fought it out to the finish' and which was used to safeguard the interests and the conditions of existence of the ruling class, that theory which conceptualized the bourgeois tenant and the peasant as a class and set them against each other while simultaneously concealing this confrontation with a general formula.) In a different case — and not only in the declining but also in the revolutionary practice of the classes — we see that theories which are false from an epistemological point of view prove to be employable in practice. (Luther and Melanchthon still insisted on the geocentric theory of the universe; as opposed to the achievements of French materialism, Robespierre and his followers were moved by an idealistic moral philosophy when in the final days of the Jacobin terror they wanted to introduce the 'dictatorship of morals'.)

Overgeneralization always tends towards a certain course, driving social action with this emphasis. It may be taken for granted that from the great variety of individual decisions a certain 'successful' average lives on, so the controlling nature of generalization already takes root on praxis level. Such maxims as are totalized in ideology will formulate this directedness in general principles. There is the theory of the physiocrats. They proceeded from the illusion that, since the quantitative difference between product sown and product harvested was perceptible to the eye in agriculture, land was the only producer of surplus value. But this perceptibility also gave the production of surplus value the semblance of exclusiveness — which was a manifest error. But this error and illusion gave rise to a historically correct inference, namely that only the land rent was to be taxed. By this means they would have freed bourgeois production of a burden, ridding it from the pressure of intolerable levies. On account of the implied consequences which were not thought out by their inventors the scope of the theorem extends much farther than its conceptual content. In immediacy it is a source of error but in its perspectives it is the source of correct results. This is so because as an ideological formula it is the unconscious understanding of a certain tendency. The reason why it could include future moves and course of action is that the theory originated where the tension between individual 'obstacles' and individual attempted solutions had occurred and accumulated and where the accumulation of the solutions or the requirements had already selected for itself a certain course to follow. Ideology is not simply a mirror of existing relations, but it is the conceptual form of collective social practice, accumulating and taking a certain direction. This conceptual form may transmit or even spur practice on, but it does not allow to be seen to what further consequences its formulae will lead because it concentrates attention only on the present. Ideology is not a deliberate deception or the subsequent and conscious rationalization of interests, the less so since it is guided by the dialectic of false consciousness, i.e. by the deceivedness of the deceiver, a practical truth which leads to a result. Though the feudalistic semblance of the bourgeois economy deceived Mirabeau, the aristocrat, this mystification was created by the spontaneous movement of the theory and not by the conscious gesture of those who invented and used it.

Ideologies become independent as soon as they are conceptually formulated. With this we have reached their other principal characteristic. In the medium of the autonomy of ideas the true relationships are necessarily altered, to assume forms which are subject to different laws. Though they are the forms of conscious perception of the conditions of existence, still they have an independent life and, moreover, this autonomy may sometimes even be opposed to the basic determinants. The strategy of the English Revolution was formulated by the religious ethos of Puritanism. In certain cases, however, this conceptual transposal caused the ideals to turn against the very practice that created them. Cromwell's soldiers, for example, mutinied against conscription, one of the principal achievements and strengths of the revolution. They presented a petition against the new institution, referring to the very liberty of conscience which Puritanism proclaimed. The independence of the conceptual forms therefore means that social praxis determines only as a tendency the 'edge of attack' of the ideas and only functionally their structure, whilst in terms of content it is relatively indifferent towards their truth.

This transposed form of presentation means that the genuine commands of social praxis are invisible in their original forms and that their true contents become consciously realized only at some later time, by means of their practical consequences. Schelsky, for example, aptly points out that the prohibition of incest originally came not from recognizing incest for what it is, i.e., from the understanding of a real social and biological law, but from the expression of a social necessity. The constraint of exogamy was the primary motive and the prohibition of incest was the negative expression of this constraint. But the form of exogamous marriage was dictated by the danger of family isolation and the need to enrich both the social division of labour and social intercourse. A marriage system which had permitted relations within a biological family group would have impeded the integration of superior and enduring, socially uniform and cooperative forms, isolating the family and depriving it of the possibility of allies and auxiliaries outside the family. [Cf. Helmut Schelsky: *Soziologie der Sexualität,* Hamburg 1960, p. 88 ff.] Here the two phases of independence are still clearly discernible. The original biological and social exigencies combine to form a certain social necessity (allies from outside the family, elimination of family isolation) as feelings of deprivation of group activity, and then these exigencies manifest themselves as customary laws in the law of taboo and marriage rules and in their primitive ideology. The genuine relationship in this law is invisible for the 'makers'. With the ideological forms achieving independence it is here that the most important illusion-creating element presents itself. In the course of collective praxis these ideological forms bring about conceptual forms whose content they jointly produce but whose conceptual form hides their origin. Reality is both included in ideology (for praxis) and not included in it (for consciousness).

Thus we have come to our fundamental issue, a decisive element of the structure of quasi-reality: in what context can we speak about the illusion-creating function of ideologies?

From the viewpoint of social ontology we are not dealing with illusions, or not, at any rate, with mere illusions, for indeed the maintaining of the power of a particular class or the 'fight to the finish' and conscious perception of its rise to power, and the means of this struggle which become material, are ideology itself. Its material criterion is success and not immediate (reflected) conformity to reality. At the same time, as we saw this in the previous chapter, it is for the most part not in conscious selection that this suitability is created. It is the lines of force of practice which indicate those among the philosophical trends of the age which suddenly become important and exciting and therefore usable — while ideologies formulate the conditions of existence of the classes, the direction of their movements and their expansibility in a different medium, so in this respect they necessarily even obscure the true relationships. This is partly due to the demands of this struggle, though for the individual participants this smoke screen is an unconscious process rather than a conscious deception, and partly to the fact that the conscious perception of the true conditions of existence begins to dawn on them only with the decline of a particular social formation; and the greater proportion of the life of ideology is filled with self-deception. All things considered, this is not an illusion-creating configuration. In the reality formulation of ideologies there are numerous elements which create illusions, but this illusory quality becomes manifest only in another dimension, in an epistemological measuring unit. It comes to light when a particular class, in the course of expanding or safeguarding its conditions of existence, wishes to awaken to consciousness of its own situation. It can do so only through its relations to the others, for its conditions of existence present themselves only in this mediate relation but it can see nothing of this mediacy. It will necessarily form a false knowledge of both its situation and itself, although it will do correctly on the basis of this false picture of reality. The epistemological standard of judgment shows ideologies as illusory pictures of reality not only for a subsequent examination. These ideologies serve in their half real and half false nature also for the man of the age when at the time of a particular crisis he feels he ought to clarify his true dimensions in the world. The error made by Robespierre and his companions was such an illusory achievement of historical consequence. They believed that what caused their ending up under the guillotine was a gesture which ran counter to the spirit of the decree on Human Rights and to the policy of the dictatorship of the morals, though these decrees had already implied the principles — tacitly and therefore without them being able to see their implications — which ordained their inevitable failure and destruction. When in the last days of the Jacobin terror they wanted to understand their situation the only thing that they could grasp as reality was this illusion. And they failed, too, because the true relationships were obscured by the very ideology they had created. It is no mere chance that Marx writes of the historical necessity of this tragedy. This illusion motivated and escorted the successfully concluded historical struggle in which the bourgeois order came into power. No other conceptual motivation would have been suitable for the task. It became an illusion only for the leaders or individuals of this struggle. Ideology is a reality in the result of historical struggles, namely for social groups to attain their existence, and it is an illusion for the individual and particular orientation of the same groups. Many a man (from prominent artists to important politicians) lost his way by taking the heroic illusions of the French Revolution for reality, but this aberration was motivated by the illusion-creating characteristic of revolutionary ideology. Yet it was exactly this deceptive 'excess', creating a 'misunderstanding' in a historical sense, which led the class as a whole to the world-historical achievement of the revolution. The reality of ideologies relates only to the practice of the entire movement of the class.

In the life of individuals ideologies already function as illusions or mainly as illusions, and so they control them.

Add to this one further important historical mission. Ideological formations, whether of legal or of a political nature, cover a historical route themselves, and with the intensifying struggle or the consolidation of power the unconscious and well-meaning start of their development is succeeded by a more enlightened representation of interests where these same principles present themselves perhaps in a more cynical form and become further refined as the rationalization of interests or as the institutionalized forms for the satisfaction of interests. Balzac, for example, demonstrates in a wide tableau how under the First Empire the well-meaning representatives of illusions were pushed into the background and succeeded by the cynical propagators of the same principles. Deception and being deceived pertain to the historical life of ideology in the same way as well-meaning misunderstanding or the acceptance of heroic ideals as reality. But regarded ontologically, deception or cynical 'fraud' (Sartre's *'mauvaise foi'*) is likewise used for fulfilling the true conditions of existence of the classes, thus conveying a material truth. It is therefore not purely a mystification. Socially deception accumulates to become the means of the rise of a class and of the subjugation of its enemies. We can see the same structure in the sphere of manipulation, but in an increased form, brought to perfection through the methods of sociology. Behind the perfected illusion of the ruling ideology there stands the reality of the self-preservation of imperialism.

Ideology of course does not wholly and completely create illusions, self-deceiving forms of thought or forms which control others through illusions. The mission of art, whose ideological function is one of its organic constituents, is precisely to reveal true motion and to confront ideological quasi-reality with individual and typical human destinies and thereby to bring to light the inconsistent path of the development of mankind in its true form. *War and Peace* starts by describing the 'medium' preceding the Napoleonic wars. It describes the quasi-reality which under the impact of these wars falls to pieces since it is unable to give guidance to individual people or old classes and the new ones. The novel ends by showing the germs of Decembrism, the genesis of a new ideological space of orientation. Russia's critical socio-historical period takes shape in this ideological sphere. The process of both the crisis of the old philosophical movement and the emergence of the new orientation of ideas is forged in the historical furnace which patriotic war meant for the different strata and types of society. Great works are not supraideological but they perpetuate the reality of the age and to this extent also initiate a new intellectual orientation. *War and Peace* contributed to the conscious perception of the fact of the continuity of revolutionary progress, the particular course of development which lasted from Decembrism to the Great October.

Ideological illusions can thus be of revelatory character and in this respect they have a revolutionary effect, and they can be the preservers of illusions which already function socially. Ontologically these illusions are measured according to the extent that they further or curb the total social reality, whereas epistemologically they are the yardstick of the individual's demand for orientation, i.e. of his 'search for reality'.

In consequence, what is reality as regards total development can also be an illusion with respect to the individual's search for reality. And indeed, one of the chief characteristics of ideologies is that they affect the way in which we see things and by this means change the aspect of the world which presents itself for the individual. In the light of ideas we are bound to have a different view of phenomena. The world takes on a different colour and we cannot perceive the perhaps disturbing but ontologically true signals. For a while this ideological picture figures for the individual as an authentic reality, showing itself to be much more real than his particular experience which refutes ideas. It can be an authentic picture of reality because it offers an experience of fulfillment compared to which the 'otherness' of individual cases or their power to refute ideas seems to be an exception. All things considered, this experience of fulfillment has some ontological elements because it is no mere error. It actually offers a successful practice for the class or society, and in comparison with this practice the particular experiences of the individual person are only contingent. And from here, from the point of view of the individual, it is this truth of society as a whole which ideology renders perceptible in representing those facts of life which refute the content of ideas as a matter of secondary importance. This is where we have to look for the finer structure of the operation of ideological illusions.

Ideologies make one see the world in a 'different' way. The thought content works as an intuition to which we unconsciously adapt the facts. While practical consciousness only accepted the forms of presentation and at best explained them later, in ideological intuition the form of presentation is produced by the way one sees things because it is to this that we subordinate individual phenomena. There are three layers in the combination of illusion and reality. Every ideology helps to discover a whole series of new relationships, thereby transcending the errors made by immediate experience. But at the same time ideology absolutizes itself by excluding a number of phenomena from the limelight with which it scans its age, thus rendering from the outset incomplete the world picture built up in this way. But we do not feel this incompleteness since even in this fragmentary state we can use ideology successfully and so we cannot help identifying it with reality itself. And at last we also assume as our premiss many thought sequences which are merely of subjective origin, the result of the self-movement of the theory, but these, too, we 'read into' the world. Ideology-produced illusions occur through elimination, projection and by rendering things invisible. In the Spanish Renaissance drama the aristocratic heroes exchanged even their most intimate secrets in the presence of their servants, because the lackeys were indeed present, but in the evaluative sense and consequently in their human equivalence they did not exist. The nobleman did not see them. The drama of the Enlightenment on the other hand rebels against this very ideology. Lessing's and Schiller's aristocratic characters fail in their attempt at these gestures, since the self-respect of the bourgeoisie and the peasantry rebels against such a view and such a scale of values. The old ideology had lost its authenticity. But as long as this ideology has been valid it applied to everybody and it functioned as an immediate way of seeing things, endowing people and things with a different quality.

But how does ideology set about to form reality? It does it with the help of totalization. Every ideological configuration wants to explain the whole of reality or the completeness of the given sphere of praxis and account for every appearance which rises in this sphere. That is why every phenomenon is at the same time a proof that the formulae are true, and what is not a proof is either not seen, not noticed, or it is an exception, an incidental circumstance which can be ignored. And in the course of totalization the true generalities will of course be mixed with those which are supposed. Quasi-reality is not simply an illusion but it is pieced together from both illusory and true relationships to form a usable picture of reality. The source of totalization is the origin of ideology. It comes about by means of the accumulation of practical solutions and the teleological notions which accompany them. But this process of accumulation vanishes once it has become independent as an actual theory. Then it will have to be used in a reverse order. What has accumulated from the real lives and aspirations of individual people now becomes the explanation of the phenomena of life. It seems to have preexisted practical reality.

The structure of presentation 'for us' changes in the same way. Phenomena are preformed not only by our situation (the position in interaction with which they develop). Their characteristic mode of existence is furnished not only by functional use or objectification and it is not determined only by the analogical form of thought. Their ultimate significance, the explanation of their 'being so' and consequently the existential essence of their phenomenal nature are offered by the ideological frames of reference. Phenomena offer themselves as *a priori* interpreted relationships. Or to be more exact, the only thing that we can see as a phenomenon is that which has in advance a certain sense. This sense as its content, giving simultaneously the form and the matter of its phenomenal nature, is lent by ideology. To quote Ágnes Heller's example, in religious ideology the ethical components of human development were understood with the help of categories which had 'upward' and 'downward' tendencies, and the individuals' actions were judged accordingly. What they did not realize was that arguably downward tendencies (such as passionateness, for example) may have perfected the 'species-being' of a particular important individuality. This ideology is succeeded later by the egoism of the Enlightenment. The falseness and, above all, the historical uselessness of the old theory was becoming evident, but from the epistemological approach there also emerged a theory that was only half true. But in the time of the dominance of both ideologies this interpretation made a moral imperative of the individual gestures such as, for example, self-sacrifice. [Cf. Ágnes Heller: *Társadalmi szerep és előítélet* (Social role and prejudice), *Op. cit.,* p. 93.] Max Weber, writing on the development of the charisma of the office, likewise refers to the role of the general approach, a role which makes individuals seem different. The charisma of the office of the Catholic Church arose at the time when it could easily happen that the personal, ethical traits of a priest were objectionable without his charismatic quality being called in question. [M. Weber: *Wirtschaft und Gesellschaft,* Tübingen 1956, Bd. II, p. 684.] Hence it is only the priestly dignity, conduct controlled by a divine mission, which becomes a phenomenon, and this also explains the personal individuality being ethically different from it. It appears in a different, illusory light. Of course Weber knows, too, that the total expansion and inexpugnability of the domination of the Church could be realized more easily and could function more smoothly on the strength of this formal universality. What was an illusion from an epistemological point of view was ontologically, i.e., with respect to social process, a condition for the consolidation of this domination. That is why we said that ideology, as an illusion or as the form of presentation of reality wrapped into illusions, exists only for individual consciousness, but here with an increased objectivity. Ideologies exceed individual forms of consciousness, they determine the motion of thoughts 'from the outside', and we see reality filtered through this medium. The emergence of an illusion is the product of an individual form of consciousness which is, however, objective in its origin. Furthermore, as an objective form of consciousness which controls society, ideology functions as reality and not as an illusion. This dialectical relation connects the epistemological and ontological 'dimensions' and the point of connection is the historical individual himself. In the fifties, under the combined pressure of ideology and manipulation of the age, the individual person really thought that he was surrounded by enemies, that there was a genuine need for these upward revised targets. But the historical result of this view and practical attitude was that there emerged a complex social configuration which combined the achievements and the very nearly fatal errors of a new society. Illusions gained control only of individuals, but on the strength of this false reality they were still able to create a real historical process.

False totalization will be exposed once the ideas lose their historical validity, when they cease to be in phase with history, i.e. when mass experience demonstrates the difference between the conceptual form of presentation and the true facts. (Think of the ideological crises of outlook which precede revolutions: what was earlier a fact which was accepted as a matter of course becomes an object first of scepticism and then of indignation, till finally there emerges a new outlook and a new system of ideas and turn against ideologies which were earlier accepted as simple reality.) Ideologies are exposed here mainly as empty illusions, though even here they are not entirely deprived of their value since every structure of ideas bears the character of a concrete class and it can unfold its historically determined content only by at the same time contributing to the enrichment of the values of the entirety of the human race. That is why the objectifications of values of declining ages continue to exist. Knightly virtues in their original forms became conventionalized until they were ridiculous and were justly treated by Cervantes as amusing quirks. But knightly honour, respect for women and self-sacrifice have nevertheless continued to exist as ethical standards.

Though ideology works with the instrument of totalization and so creates the authentic picture of reality of the age, this totalization is itself neither false nor true. K. Kosik is thus wrong when he sees the source of illusions in the operation of totalization, i.e. in ideological thinking. [K. Kosik: *Die Dialektik des Konkreten.*] Kosik attempts to confront concrete totality with false totalization, but his arguments are sufficient only for a formal phenomenological analysis, i.e. for investigating the relation between part and whole. But the totalization of ideology measures the totality of the global movement of history against the conceptual totalization of a conceptual and ideal synthesis. This means only that every element of reality derives its meaning and its explanation from a determinate conceptual centre, but this centre faces in the direction of the historical movement. Epistemologically this totalization, too, is 'false' but, since on another plane it is identical with the trend of historical development (it makes this trend and also embodies it), it turns into a correct and valid truth. Only history can be the yardstick of the falseness of totalization. It is false when the elements of reality become subordinated to a conceptual unit whose direction and motive power cannot establish contact with the concrete struggles of history. Its illusoriness already shows through here and it is therefore also useless.

This is where we can see the difference between the rationalization of interests after the fact and the ideological forms. Though these two spheres are in contact with each other, they are nevertheless separated by distinctive boundaries. The rationalization of interests is an immediate manifestation of practical relations. It has not the surplus of conceptual invention, i.e. of the nucleus of ideas which points in the direction of 'species-being'. Here the theory does not have an independent and heuristic function. While ideologies boldly put aside the ready-made forms of presentation, offering perhaps illusory solutions or in any case ones which are in contradistinction with the existing visible sphere of phenomena, here direct manifestations of the forms of presentation are emerging. From the already existing material of theses which has been refined into a cliché, the rationalization of interests raises 'panels' to explain or mystify a particular situation of interests. These 'panels' form a structure which is somehow in harmony with the structure of the forms of presentation. It expresses the system of an already functioning course of action.[12]

[12] The following pages give one possible interpretation of Marx's category of ideology. I presume that in Marx's oeuvre the term 'ideology' figures in several senses, among others in the sense of the rationalization of direct and immediate interests. At the same time he lays particularly great emphasis on explaining ideology's characteristic of creating reality, the ontological category of ideology which Lukács reconstructed in his *Ontology*. I take this conception as my starting-point, although later when I examine one component of this illusory sphere I shall depart from it in order to focus my attention on the analysis of the subjectivistic role of this sphere in concealing reality and functioning as an artifical reality.

One of the functions of ideologies is likewise to stimulate functioning social praxis and to enable it to justify itself. But the only way in which they can fill this role is if they conceptually generalize their task and in doing so go beyond their immediate and practical goals. That is why there is always more in important ideological achievements than what the age can consciously use of them. It is only posterity which will discover and turn to good account a large proportion of their content. Remember, for example, the difference between the contemporaneous and the ensuing effects of Spinoza's philosophy, or the after-life of the theory of natural law which exerted its maximum influence several hundred years after its origination. Generality is a criterion of applicability, but it is at the same time an instrument of intellectual pioneering. That is why ideological configurations mean more than the immediate expression of interests in the nature of a 'shopkeeper mentality' (Marx). (Robespierre, as an ideologist, went beyond immediate class interests just as Ricardo did, and the great works of art express the total motion of an age for the historical entirety of a class — from the level of motion in the sense of human development — just as the important philosophical systems are also more than only manifestations of immediate class interests.) Though our entire conception of ideology is based on emphasizing this difference, there is nevertheless no harm in once again recalling the way in which Marx drew the dividing lines between the two different configurations. Vulgar economics endeavours to give facts a 'pleasant' interpretation, that is, it tries to rationalize interests, whereas our ancestors, the classical economists, advocated total motion which gave the class a stimulus — perhaps even against their immediate interests.[13] He makes it unmistakably clear that the ideologist expresses class interests only so far as they represent the given forms of human development. Vulgar economics or the rationalization of interests, however, strives only

[13] Whilst the form of alienation keeps classical and therefore critical economists busy and they attempt to peel off the onion-skins of this form by means of analysis, vulgar economics feels quite at home in precisely alienation. [Cf. *Theories of Surplus Value*.] And against this shallow conception, he writes about Ricardo as follows: Ricardo's relentlessness was thus not only scientifically honest but also scientifically binding for his standpoint. Consequently it makes no difference for him whether the further development of the forces of production destroys landed property or the worker. If this progress renders worthless the capital of the industrial bourgeoisie, he welcomes that in the same way. If Ricardo's conception is entirely in the interest of the industrial bourgeoisie it is so only because and so far as its interests coincide with the interests of production or of the productive development of human labour. Where they conflict with them Ricardo is just as relentless towards the bourgeoisie as he is otherwise towards the proletariat and aristocracy.

for this particular assertion of interests, in opposition to the general tendency and consequently the general interest of the development of mankind.

The impelling tension of ideology or the alerting power of conceptual invention is thus the tension between class-relatedness and 'species-being'. For indeed, it is this class-related form of presentation and expression which grasps people — perhaps by 'deceiving' them — and makes them the figures who are moved by their history and at the same time the creators of this same history. The final source of illusions is therefore the transpositional and necessary form of presentation of 'species-being'. Only thus can the individual person have a say and be a participant in the total process.

Here light is shed on the fact that the capacity of ideology to create illusions, i.e. its quasi-reality, is the consequence of a historical process of alienation. If people could create their history immediately, there would be no need for false consciousness, the roundabout way of ideological illusion which transforms the outlook of individuals. But what prevents this immediate shaping of history is the multitude of social transposals between individuality and 'species-being', the internal articulation of classes, nations and group antagonisms, or in short the alienated, class-related nature of immediacy and within it its exclusive reference to the private sphere. Ideology serves as means to overcome this inconsistency. It guides man's 'species-being' and his mission towards mankind as a whole, but only through the transposals of the class-related conditions of existence and the illusions which these conditions entail. Ideology assumes an independent form as a seemingly unattached configuration because over against it man's progress as regards his 'species-being' also becomes independent as the advantage, the life-enjoying role, of class-related or privileged groups. This is why Marx writes, as we have already mentioned, that in the beginning this development of the abilities of the human race goes on to the detriment of the major proportion of human individuals and even of human classes, but it finally breaks through this antagonism and coincides with the development of the single individual. That is, the superior development of individuality can be bought only at the expense of a historical process in which individuals are sacrificed, since the advantages of the species always break through to the detriment of the advantages of the individuals, since the advantages of the species always coincide with those of specific individuals. [Cf. *Theories of Surplus Value.*] Ideologies express this duplication, the separation of 'species-being' and class-relatedness, individual nature and species nature, their mutually opposing movements in terms that also nevertheless give expression to the progress of 'species-being'. The alienation of thought and its gaining independence with respect to practical reality is only an outgrowth and an expression of the outward separation of the individual (and of class) from the species.

Conceived in philosophical terms, the quasi-reality of ideology, its illusoriness as well as its ontological correctness stem from the dialectic of 'being in itself' and 'being for us' of social existence, a dialectic which bears no resemblance to the relation of *an sich—für uns* in nature. Natural 'being in itself' which human praxis and consciousness attempt to identify is invariable even in the long term. It is changed at best by the entirety of society but only as regards the conditions of its existence. (With its deforestations, for example, society creates a new meteorological medium for itself but it still will not have changed the laws of nature.) Whatever this interaction may be, nature only allows itself to be shaped and used, and man approaches its forces and its phenomena 'from without'. In society, however, the subject creates processes from within. The individual is not only the creator of the laws of society but he is also determined by them. Social 'being in itself' also includes man's subject-being, both as creator and as created, whilst natural 'being in itself' excludes it. It can be approached only 'from without'.

This also means that the consciousness of an age also belongs to the social 'being in itself', i.e., to its objective essential processes. This consciousness contributes to the development of this process in the same way as does the praxis which it controls only to a certain extent. It belongs to it because in decisions and individual actions it changes into material effects and causal chains. Men's actions are directed not only by the objective forces which are unknown even to them but also by their conscious decisions. They cannot select the conditions of existence of the age since they receive them ready-made, but these conditions are the results of the half-conscious, conscious or unconscious practice of earlier generations. The element of consciousness therefore cannot be excluded from social 'being in itself', though the ontological datum of social conditions and our understanding of it (its 'being for us') are, after all, separated from each other. But the structure of unity and difference is other than in nature. The forms of consciousness which have become public property in society (ideologies) also function as objective constituents (norms, motives), and the individual person, in understanding the world as it is 'for him', reflects both what follows from the objectivity of the conditions of existence (as much as he can grasp of this) and what he has received from the ideological medium of the age. 'Being for us' is not simply the form of presentation of an objective datum, but it is also shaped by the world of ideas of the age and by society's understanding and making this world. That is why we cannot draw a sharp dividing line between the truth of ideologies in the ontological sense and their epistemological truth. It is true that they are separated from each other but this difference always remains relative. In 'being for us' there is always left some subjective element of consciousness, distorting momentum or illusory 'phenomenon' and 'being in itself' is likewise formed of many objectified ideological elements.

The process of alienation of the ideological sphere can be observed in every historical period. It is part of the process by which a class as 'it is in itself' becomes a class 'for itself', i.e., when people become aware, even if on the level of false consciousness, of the demands of their common existence. The naive spontaneous use of ideological means gives way to the reflective and then manipulative application at a time when a sphere of ideas, which has been glimmering in interests and in the requirements of practical measures, assumes a fairly distinct outline and must be rationalized or invested with values. (This is the process to which Engels refers when he writes on the change of the naive puritanical ideological outlook of the English Revolution: Locke replaces the arguments of the Old Testament by the 'glorious revolution'.) But with the rise of particularity and the development of bourgeois society this process of alienation is consolidated so as to surpass in strength all earlier formations. Since the Enlightenment there has been hardly a break in the process in which ideology — now almost embracing the entirety of the conceptual legacy of mankind — becomes a useful, independently functioning system of arguments which can serve rationalization. Of course, this process of alienation manifests itself with different characteristics in different ideologies. Manipulative tendencies do not exert their effect without mediation. In politics the tendentious role of alienation is stronger, whereas in arts we can more strongly sense the effect of controls, endeavours aimed at awakening man to consciousness of his true nature and his role in society, the need to free him from manipulation and alienation. But we shall not concern ourselves with examining the transitions between these poles, i.e., with the peculiarities of particular ideological configurations.

The historical background to the change of ideological outlook is that bourgeois society tears apart the profane and ideal forms of self-expression inherent in the strongest mode of existence. This is where we find the strongest contrast and relationship between material practice and the world of ideals. And it is as early as the French Revolution that a tendency gets under way which aims to exploit this historical fact in a manipulative manner. It tries to govern with the help of ideas.

With the invention of historicism the historians of the Restoration period also discover the possibility of treating ideological legacy in a manipulative way. Besides present requirements it is possible to raise formulae of rationalization from the storehouse of arguments of the past. And moreover, polemic on the dictatorship of morals also gives a glimmer of this alienated and manipulative use of ideological arguments. The origin of this tendency lies in the idealization of *citoyen* existence and in its conceptual verisimilitude, even if Robespierre and his associates covered this route with tragic good faith. It lies in the duplication on which Marx wrote in his fourth Feuerbach-thesis and which is the fundamental phenomenon and the basis of the entire bourgeois development, namely the contradiction between the reality of the profane way of life of the bourgeoisie and the ideality, the unreality, of communal and social existence. The emerging bourgeois order then profanes this dichotomy. Hypocrisy, feigned morals and the phenomenon of rhetorical ideas become an average fact, and the part of ideology which is taken into everyday life becomes indeed a form of rationalization and deception. (The Opium War is fought in the name of the Enlightenment and Christianity, and for the average European citizen of the age it manifests itself as a natural spiritual necessity, whereas the bloody brutality of gross interests hardly shows.) Marx's theory of ideology represented a discovery with respect to two directions. On the one hand, in opposition to the alienated practice of manipulative application it revealed the ontological background of the general forms of ideology, the material necessity of their origin behind the arbitrary application, and by discovering the precedence of the modes of existence, i.e. the reflective role of ideology, his theory traced back the alienated configuration of thought to its historical origin. And on the other hand, it embedded this sphere in the reality of practice, in the logic of recognizing the conditions of existence and fighting them out to the finish, where the tendency of independence could also be included. The theories of Social Democracy and dogmatism dissolved this integral whole, and ideology changed once again into a mere rationalization, into the justification of the arbitrary stressing of quotations and tactical manoeuvres. The functional application of Marxism was born not in the process of Stalin's pragmatic handling of ideology, but in the simplification made by the thinkers of the

Second International, namely that ideology is the concealed conceptual form of interests and is created with an intent to deceive. Hence it follows that Marxism can also be used as this kind of tactical weapon in the present of tactical struggles. For any measure at all it is possible to find a thesis which can justify even contradictory moves. That is why the Marxist method and indeed the legacy of Marxism recedes into the background and becomes at best the collection of quotations. Ideology became here a means for the arbitrary creation of illusions. The intention which has just been decided, which still exists only as a plan, after receiving regular confirmation, but of course before it is realized, became a reality in people's minds, indeed a more real one than the facts of their own lives. That is the way ideology took the form of a devout attitude in everyday life, the external marks of which bore a formidable resemblance to the religious attitude. In comparison to the primary truth of ideological theses the real tendencies sank to the degree of secondary data with respect to which a man who accepted facts could feel only a consciousness of guilt for having given credence to a seen and experienced world. This ideological guilty conscience is one of the psychological explanations for the course taken by the trials. The involuntary sense of guilt was created, over and above the revoking of the entitlement to moral defence, by the illusion that a particular tactical move was the necessary consequence of the entire theory and the entire revolutionary praxis and that at the same time it seemed a more real fact than the experience and conviction of the individuals. To deny these 'facts' or to simply ignore them was tantamount to the renunciation of the revolutionary theory. The individual could confront ideologized reality only in his particularity and at the same time he was called to account for collective theses as unnoticed or denied elements of reality. And the artificial retention of particularity gave rise not only to illusions but, in more serious instances, also to a guilty conscience. It was the everydayness of guilty conscience which made the ideology of the period, coming to an end with Stalin's death, into something which was more real than reality itself. All this, however, is psychology and largely belongs to the problems of false consciousness. It demonstrates, however, the regenerative capacity of Marxism, namely that even during the time of the grossest distortion it could maintain its sense of reality, its resistance against manipulative methods and thereby the continuity of its progressivity, even when in the Stalin era this could manifest itself only in latent, even false, forms of consciousness. The new era initiated by the 20th Soviet Party Congress (1956) would have been inconceivable without this internal vitality.

3.6. FALSE CONSCIOUSNESS AND MANIPULATION

False consciousness is also a constituent, a phenomenon of historical praxis. It is one form of the difference and the unity between the 'made' and 'consciously made' individual way of life and social practice. Or in other words, it is the contradictory unity of action and consciousness of action. We already pointed out this difference when we discussed the structure of ideologies. They separate as forms of consciousness from everyday life, though they grow out of it and form its constituents. Is there any difference then between these two configurations of consciousness? There is one indeed, namely in the way they relate to action. False consciousness is the conscious vehicle of motives of an age, it is the arsenal from which people take their incentives and teleological notions for their actions whose content, direction and structure considerably differ from these conscious intentions and motives. In contrast to this, ideology serves primarily for the understanding of reality, the rationalization of already accomplished deeds or the final, theoretical explanation of intentions. The difference between the two media is relative but nevertheless essential. In ideologies it is reality which is measured against a theoretical system and becomes a structure of ideals or norms. In false consciousness this motion is reversed: historical praxis picks from the abstract 'thesis material' of the ideologies whatever it happens to need as a motive with a capacity to control actions. Ideology, therefore, contains the world of pure conceptual forms because it is the abstraction of praxis, whereas false consciousness is the medium of 'pure practice' since it is the practical abstraction of a theory which has been turned into a motive.

So our actions are based on false consciousness, and since we do not get anywhere with the pure conceptual forms of this configuration of consciousness we must translate the abstract theses into usable motives. Yet this 'translation' is rarely conscious. It is mostly created spontaneously in the 'unconscious' team-play of the entirety of society or a particular integration where this team-play transcends the individual. Man knows what he 'does' (i.e., the kind of conditions and historical consequences that his actions have) just as little as he can realize how he inherits this system of motives which directs the sphere of action or to what extent this system gives him an adequate picture of what he does.

In order to understand the difference and the unity of ideology and false consciousness let us select a classic historical model. All branches of the German Reformation were significantly influenced by the ideologies of heretical movements and especially their conceptual syntheses, the works of Meister Eckhart and Heinrich Seuse of the Dominican Order. The ideological basis was furnished by the discovery that man became identical with God immediately and that this identification had no need for the hierarchy of the Church and, least of all, for social and spiritual subjection and servility. But the trouble was that this thesis manifested itself not only in an excessively abstract form, in the manner of the equivocal and obscure theses of mysticism, but also that it was incomprehensible for the mass movement. Johann Tauler and Münzer, who further developed and vulgarized Eckhart's idea, endeavoured therefore not only to reveal the useful content of this system of thoughts but also to reshape these theses in accordance with the needs of the mass movement. What with Eckhart was still the mystical identity of man and God changed with Tauler and Münzer into an anguished process of identification full of sufferings. Man cannot immediately be identified with his God. He must fight this out for himself. He has to undergo the Calvary of this journey insofar as he has to destroy the social obstacles which are in the way of this identity. It is true that man chooses his God in an autonomous way, but the only way in which he can attain this autonomy is through a struggle with himself and the world. What with Eckhart was the peace of the soul became with Tauler and Münzer a war between the body and the soul. It is thus that it could become the system of motives and the false consciousness of the peasant revolution. The original ideology underwent a transformation because it acquired a different life under practical needs.

This difference is even more obvious in the practical application of the theory of predestination elaborated by Calvin and the conception of predestination professed and practised by Protestant sects. Calvin deemed it impossible that man could make certain of being chosen by God for eternal life. Predestination, as a theory, disregarded the individual person. The elect live their lives in just the same way as other people. It is only after death that they get to a higher status. But the important thing in the everyday struggles of Protestant sects was precisely the self-assurance of the individual, that perceptible 'surplus' which verified their separation and their qualitative superiority over other classes. They wanted to know why and in what respect their way of life represented a moral and ideological category which was superior. With this end in view, religious ethic and dogmatism transformed Calvin's conception. In the success of his work the individual man can surmise the divine inspiration. The success of one's financial activity can afford proof of being among the elect and, in this respect, direct one's activity day after day. Business success is the hallmark of *certitudo salutis,* whereas failure indicates the loss of the state of grace. So there is some sort of a context which can guide the individual in his everyday life. Predestination has remained as a historically controlling, cohesive ideology, but it underwent a change in practical 'making': it became a false consciousness of primitive bourgeois societies as the everyday solace and comfort of the elect.

We have enlarged on these examples in order to distinguish ideological forms more clearly from false consciousness. Though this difference, as we saw, is relative and the connections are fairly important, we still cannot identify these two concepts with each other. We would like to stress the difference because bourgeois society, especially since Karl Mannheim, has handled these two categories as synonymous forms in the same way as dogmatism. According to Mannheim, the 'falseness' of false consciousness indicates the difference as ideology and utopia lag behind or, with their fantastic concepts, run ahead of the reflection of the real elements of history. Correct political and historical action is then hindered by this lagging behind or running ahead.[14] Ideology and false consciousness are identical concepts in that both serve to denote an extrahistoricity, but false consciousness expresses more clearly the action-alien character and the error. (Mannheim's theory is symptomatic as it indicates that the identification or differentiation of these two concepts is not a sterile conceptual debate but a function of the problem of attempting to understand the course of history from the decisions of the individual people, from the accumulation or the alienated autonomy of their erroneously true notions. The moment we see only an error or a distorted reflection in the falseness of consciousness history becomes independent, breaking away from the decisions of individual people, and the contradictory relationships which function between individual activity, i.e. teleological and conscious action, and the forming of history will no longer be obvious.)

Dogmatism, under the influence of different intellectual motives, follows the same road. Just like bourgeois sociology it, too, sees only

[14] Consciousness rates as false if it gathers information on the basis of norms by which we cannot act even with the best of intentions (a certain condition of existence), where we must not regard the individuals' failure as an individual failure but as an action under the pressure of a falsely conceived moral axiom. Considering it from this point of view a consciousness whose method disregards new reality is false and ideological. [Cf. K. Mannheim: *Ideologie und Utopie,* Bonn 1929, pp. 51 and 53.]

error in false consciousness and identifies this self-deception with the ideological forms. This identification is also furthered by those early formulations of Engels in which he makes no reference to the separation of these two spheres. But at the end of his life Engels already corrects this conceptual identification, for example, in his renowned letter to Mehring (14 July 1893). The reason why these two concepts are again identified with each other in the subsequent development of Marxist theory, in the ideological debates of the Second International and then in Stalin's practice, is that ideas in all their forms became instruments of manipulation, and politico-economic subjectivism could thus gain a theoretical support as well. Ideas, whether they are true or false, serve the purpose of directing people. Ideology inevitably became identified with thoughts which were either alien to reality in terms of their content or false, but which still had the capacity for stirring people.

But let us come back to the basic question of false consciousness. Its structural nature lies in its offering and using a conceptual system of motives whose thought content gives indeed a false reflection of the concrete world but as a motive leads to correct action. It homogenizes the actions of classes and integrations by knowing something else about the world than what it does with the help of this knowledge, and these two different media nevertheless become identical in the result. Or to put it more simply, in the 'knowledge of making' the world is reflected differently from the way in which this consciousness is formed as making, and with the help of this consciousness the world is formed as making, and notwithstanding this difference turns the practice of society in the right direction. To understand this difference and this unity, think of Max Weber's famous example, the ethic of Puritanism, the discovery of the 'asceticism of the world' and the relation of the development of capitalistic practice. Weber has a striking formulation of this duality: Growing rich was with the Puritans merely the unintentional consequence of their virtue, although it was at the same time an important mark of that virtue. It was only the rational ethic of Puritans who focussed their attention on the next world that led to the economic rationalization of this world precisely because nothing was further from their minds than getting rich, precisely since the work of this world was for them only the expression of an ambition for a transcendental goal. [Cf. M. Weber: *Gesammelte Aufsätze zur Religionssoziologie,* Tübingen 1947, Bd. I, pp. 532 and 534.] Weber compares the explosiveness of this dual motivation with the thought system of those oriental religions which lack any such dynamics, and that is why he emphasizes the contrast between known goals and the enrichment which ensues because of them but not intentionally. The 'known' and the 'made' confront each other without mediation and they meet only in the result. This contrast is also true regarding the history of economics since in reality capitalism begins not at the stage where immediate profit, i.e., the need for getting rich at once, is the goal and consequently where profit is made within the circulation of money but starts beyond all these factors, in the sphere of production which premised a previously unknown mediate relation to the financial resources, individually and socially alike. Individually because the returns of industrial capital could no longer only be 'expected' as in the case of monetary capital but its stages had to be created day after day. And socially in the sense that there came about a universalization of the division of labour in which an unending sequence of mediations were placed between the 'makers' and the accomplished results returning to them.

The falseness of false consciousness is therefore contained only in the immediate sense of the motives. The impulses to action which they convey and the directions for action which they include are real. (This assertion refers to the main trends of history, for indeed there are also forms of consciousness which because of their overdevelopment or prematurity will be false in their results, too. We will come back to that later.)

But how is it possible to come to the correct result from false motives? The answer is this: socially it is the correct solution (or its germinal form) which first comes into existence, but social consciousness can recognize this tendency only in the form of the available conceptual motives which lead immediately to this result. It does not see the process of true development but only the desired or unacceptable contours of the result, and it wants to 'make' and understand this together with motives which lead to this result but which do not originate in the process of development itself. Here, too, the logic of final causality is at work. Whilst in reality there is an independent result created from the many different kinds of individual motives, a result which differs from all particular efforts, in consciousness this relation is reversed and the result attracts those intellectual motivations which may — virtually — lead the deeds there. The subject of the religious controversies of the 13th—14th centuries is the autonomous or 'externally controlled' form of salvation: is it in the power of man who is left to himself to free himself of his sins, or is it only by depending on God and Church that he can enter into the state of grace? But opinions are divided. One camp, following Saint Augustine's theory of predestination, thinks in terms of the practice and need of conformity to the early feudal order, and so in accordance with the principles which teach people to accept acquiescence to the hierarchical mode of existence. The other camp, through the gnostics and the Albigensians to Fiore, Tauler and Luther, defends the independent ability of the individual to deliver himself. In a recondite form even for the debaters themselves the dispute implicitly contains the motive which leads to social liberation, to the way of life of a different nature. But the 'already made', i.e., germinally existent, way of life manifests itself as the unconscious element of 'inclusion' of this immediate motive. The motive is 'false' because, instead of reflecting the causal sequence of the real processes, it is substituted into this series of motives 'from without' as a teleological cause. But for the individual it seems that things follow the *primus movens* of the thought. It is at the same time also obvious that a particular social tendency can grow only if more and more people identify themselves with these ideal motivations and perhaps even give their lives for them.

On the other hand, there is no denying that false consciousness is not only an illusory motivation of true relationships. Its history abounds with elements which mislead in the results and which are also 'false' ontologically, with illusions which do not contain, even as an 'inclusion', a true element that can be verified later in the course of history. This can be observed mainly at the time when a particular formation comes into being; the great variety of possibilities of 'what will be' still crowd together in the manifoldness of the way of life which is to be worked out experimentally and therefore the conceptual formulation of this manifold possibility is confused. Falseness, as a framework of synthesis, also follows from the uncertainty factor of directedness to the future since both experimenting practice and false consciousness can orientate themselves only in the sphere of possibility. So there is a double filter at work: imagined, hoped for and desired possibilities are interlaced with real ones, whereas these possibilities, as data, are absolutized as reality for conceptual and emotional formulation. The average consciousness of the age does not select. It concentrates not on the elements which posterity will realize but on the entirety of the web of ideas and illusions, desires and true elements. And it may also occur that certain groups are connected only with the outward aspect of the layers of illusions which they want to carry through despite the real content, and their failure is thus necessary. Hence false consciousness does not always and, what is more important, not immediately in terms of time yield some result, and it deceives not only individuals. Its truth is realized after the example of this failure, often through the downfall or sometimes the destruction of whole groups. The failures enforce the perfection of these conceptual forms.

Being deceived, or even the destruction which follows false motives, thus pertain to the historical existence of this form of consciousness. We have emphasized purposely that this concerns groups: classes cannot be deceived for the simple reason that the blunders become corrected in the average of total movement. The passable and impassable paths are tested only in the process of becoming a class, i.e., in the development from 'being as it is' to the class 'as it is for itself', and this may also mean that those who stake their future on a blind alley will perish. Learning from their example the entire class can move in a more correct direction. (From the sectarian movements to the Reformation one can see this dialectical evolution. It was through relentless deception and destruction that false consciousness had developed here the historical form of consciousness which then became suitable for directing a real social transformation. But this process of development is several hundred years old, and it was not the immanent reality of false consciousness but its illusoriness and its nature of being a pitfall which were dominant during that time. People make their history by luring themselves into a trap through their collective existence, a trap which will be forced open by the oncoming generations.) The ontological truth of false consciousness emerges not in the individual man's praxis but in the average of the social evolution of a lengthy historical period, in the balance of many individual impasses and successes, and in the selection of assumptions with ruinous consequences in terms of practice and of theories which make correct approximations. That is why the deception of the individual or the group and, with it, the sphere of the subjective illusory world cannot be detached from false consciousness.

Allow me after this to hazard a slightly astounding thesis: as regards the individual person's sense of orientation and the picture of reality which he forms of the present, the 'falseness' of false consciousness is relatively independent of the historical level of consciousness. That is to say, the individual person in the present of every age necessarily orientates himself in 'false', illusory bodies of thought. It is true indeed that the historical process of the enlightenment of mankind can more and more eliminate errors and that rationalism gets increasingly nearer to the 'core' of reality, nor do we wish to draw an agnostic picture of society. But this uninterrupted development of the self-knowledge of society is true only as regards understanding after the event. The scope of action of the existent present is open towards the future and, due to the entangled relations, orientation with regard to the future is increasingly complicated. The real orientation is formed by the whole of the social movement which we can consciously measure only after the event. In the present we can only approach this orientation since our theories can comprehend only fragments of this movement.

It was indeed the 'great adventure' of the past sixty years of socialist development which called attention to this experience. The most rational theory in world history came nearer than any other theory to understanding social processes, but it still functioned as a false form of consciousness regarding the final tendencies of the structure of the new order. It is not primarily to the tricks of manipulation that we refer here, still less to the change of consciousness of those figuring in the trials or to the error of 'I believed it'. We allude here mainly to the scientific and yet 'false' grasping of essential political tendencies and fundamental laws of economics. (Remember how the economic laws of the socialist development of industry changed, how little it could be seen that the primary task of socialist industrialization was to overtake the capitalist economy. And while it did carry out this task it did it with a 'false' scientific motivation.) Science therefore does not exclude the possibility of false consciousness taking effect. That is why we regard as absurd the rigid confrontation of ideology with science and the diagnosis that the present time is 'disideologized', that false consciousness has gone.

Manipulation takes hold of man likewise by means of false consciousness. Whereas ideological configuration wants to raise him beyond his particularity by always demanding something from him, the alliance of false consciousness and manipulation fixes him in his particular dispositions: let him be the way he is or if he wants to change his life then let him choose below his possibilities. It is through false consciousness that manipulation takes advantage of man's self-knowledge and grasps him while at the same time concealing the true state of affairs from him. Secrecy and self-deception are the poles of manipulation. Vance Packard describes manipulated consciousness which has already become a style of life as one of the essential tendencies of the American way of life. According to him, every social stratum and every man therein wants to seem one class better than he is and he also believes this pretence which he has created himself. It is in accordance with this false consciousness that he furnishes his home, dresses or chooses his car. It is this illusion that he presents 'inwardly' and 'outwardly' alike. [Cf. Packard: *The Status Seekers,* New York 1959.]

As a compensation for illusory mobility, people obtain the peace of their particular way of life and the fact that they do not have to change their lives, that they can identify themselves not only with the false illusions that they have created of themselves but also with their morbid, alienated way of life. But the fiendish thing in this attitude and lack of self-reliance is that people believe these forms to be their self-realization since they cannot see that this 'wanting to seem more' is dictated by the production system of capitalism.

Manipulation, as we have said, functions by means of secrecy. Neither is this a radically new thing. The unequal 'distribution of secrets' between rulers and subjects characterizes the structure of every society. The order of secret measures, agreements, alliances and decisions is a weapon of every leadership apparatus: knowledge is power and ignorance is one of the preconditions of subjection. But modern capitalism has extended this method to private life and invested it with a semblance of publicity. There is an invisible apparatus which controls forms of conduct from dress to the uniformity of thoughts, and all this gives the illusion of the freedom and autonomy of individual choice.

A fundamental problem of socialist development has therefore from the outset been the creation of a democratic publicity. It is no accident that the *State and Revolution* attached such great importance to the democratic 'opening out' of political 'secrecy'. And even if this turned out to be unrealizable in the transition period, without it it is impossible in the long run to avoid bureaucratic and manipulative forms of development. It was the Stalin era which built up the web of bureaucratic secrecy and the hierarchy of secrets, attributing a leading role to tactics as opposed to strategy and theory. It follows from this principle that tactics became the model for the control of society. And one of the basic tricks of tactics is to keep certain measures secret, to control publicity, to divert attention and to make spectacular propaganda for artificially created 'facts'. It is invisible manipulation which shapes these visible 'facts', but it is a theory accepted by everybody which bears them out or makes their acceptance morally binding. Here secrecy is a means of 'gross manipulation', a weapon for the tactical deployment of artificially created 'model-facts'. Like the drama of the trumped-up trials or the dualism of the Party's organizational structure — which was seemingly selected from the rank and file but was controlled in fact from above — precedent, the particular case, or circumstances detached from their context corroborated this artificial, manipulated factuality.

In dogmatism, too, the particular individual is left to himself. He confronts the visible and isolated facts only with his abstract conviction. If he accepts them he will have to modify his theoretical conviction and if he denies them he will clash with the entire process of socialist construction. In this way the manipulative practice of dogmatism shapes the new form of false consciousness, an overall collusion between manipulated facts and the one-sided development of political consciousness. Consciousness invests the forms of presentation with a different scale of values and existence. It repaints things since it sees them in a different order of quality. There will be a change in the hierarchy, 'essentializing' and classification of phenomena so that the important and the insignificant, the decisive and the secondary, become differentiated already in the way they are perceived. But this differentiation occurs not according to the experienced frequency, statistical proportion or usual context but in compliance with the organizational principle of the new thesis. Conviction will deem ordinary cases to be contingent and therefore secondary and exceptional factors to be important. (The reason given at one time for the shortage of bread was that 'kulaks' had been buying it up. It is true indeed that, owing to the bread shortage in the villages, we could see peasants who were buying bread in Budapest and this contingent fact manifested itself as a law subsumed under a proposition which was alive as a conviction, and neither the true cause nor the genuine content of 'hoarding' as related to a class was evident.) That is why we said earlier that ideology and with it false consciousness do not simply create a new organization of values. They do not merely repaint otherwise identical phenomena with the categories good and bad, useful and useless, beautiful and ugly, but they penetrate into their essential quality. They rearticulate the phenomenal world. For example, in the forms of conviction of false consciousness even the 'signals' which sometimes issued from behind the scenes of manipulation became invisible. Not only did the blunders made in the trial procedures vanish, such as the 'fact' that it was impossible to break the resistance of Traicho Kostov, but also the factual contradiction between the irrational directives coming from upward revised targets and the 'answers' provided by the industrial structure. As false consciousness, the 'principle', conviction, could be more genuine than reality itself because it was also supported by some 'facts', because false consciousness builds its ontic construction on self-verifying manipulated facts. What verifies a proposition which is alive in conviction is the more real reality. What contradicts it does not exist or is a reality of secondary importance. Conviction is of course also the consciousness and the accepted motives of true relationships. But in its configuration which is compelled to particularity it is relatively indifferent to the existence of facts because it also creates this hierarchy of existence in an autonomous way. That is why we do not simply talk about belief here. Belief develops the scope of conviction and of the subjective picture of reality, independently of facts. The more the facts contradict it, the more it sticks to its principles. The principle of *'credo quia absurdum est'* furnishes the substance of this attitude whereas false consciousness, as a conviction, grows out of the complicated web of principles, conditions of existence and everyday praxis. Its source is a structured fraction of the facts and consequently a part of reality. The tragic crime of dogmatism was that it built up a self-deception which was based on fragmentary factuality. The Marxist postulate became obscured, namely the postulate which deemed that for the ideologist and the revolutionary the essential requirement was to explore the 'cunning of history', that is to say the true hierarchy of facts, even if this new web of facts clashed with all earlier theses. The factuality of dogmatism was orientated by an abstract 'ghetto of facts' and made this attitude the basic requirement of party-mindedness, i.e., to start not from the whole, from the totality of facts and relationships, but from the privileged and processed facts which decisions and directives have already made suitable for substantiating the rigid formulae. In such an intellectual climate, to notice certain signals which contradicted an official valuation or directive simply rated as a contravention of the proprieties. And this contravention gradually became a fetishized form of conviction. True conviction, besides becoming identical with blind acceptance of the valuation received from above, also brought about an internal manipulation in which the individual simply considered himself guilty if by any chance he noticed such discrepant signals. Still later, he even gave up noticing at all. The manipulative intensification of false consciousness formed a quasi-blindness which simply concealed reality from those who played a role in the life of the community.

In the forms of the aberrations of false consciousness there comes into being a definitively subjectified form of illusion. Each day the world changes its forms in accordance with tactical requirements but in a way invisible to the individual who, enclosed in isolated or artificial integrations, is unable to distinguish true totality from factitious wholeness. He takes manipulated facts for real and completes these facts with the theory he has. False fact and false consciousness find each other. At this stage of the fetishizing of consciousness there will be a multiplication of the transposals through which the factual relationships of reality can still make their effects felt. That is why in the socialist economic and political development it is more difficult to detect and correct the wrong tendencies in due time.

But the fetishization of consciousness can never be complete, especially not on the ground of the revolutionary ideology of Marxism. Although the manipulation of the Stalin era brought about a short-circuit between the particular man and the particular fact, a nearly complete suspension of individual autonomy and a changing into 'secrets' of the facts which manifested themselves in the great relationships, false consciousness nevertheless did not work in the same way as in earlier social formations. There came into being a peculiar 'background consciousness', either in the *modus vivendi* of cynicism or in the form of scepticism, suspicion or suppressed rational doubt. By the concept 'background consciousness' we mean the phenomenon that individual people suspected in the depth of their consciousness that 'there was something amiss here', i.e. the paradoxical mode of conduct which consisted in accepting illusions by dint of using their conceptual discipline to oust the disturbing 'signals' from their consciousness. Ehrenburg's memoir (though it projects the moral enlightenment of the present on to the forms of conduct of the end of the thirties) has an excellent observation of one reaction: when someone's closest friend was arrested, the obvious refutation sank almost automatically to the depth of his consciousness, and the machinery of self-persuasion set into motion with the known refutations vanishing into thin air. They only got into the background. Then some new and disturbing facts got into his 'background consciousness', facts which he was not supposed even to think about, not so much because of the 'outside' danger but because of the instability of the inside defence of his loyalty and his conviction. But the trouble was that the more the disturbing facts were deposited in 'background consciousness', the greater (although unconscious) the tension it represented. What helped one most easily to get through this mass of refutations was cynicism which, however, gave rise to endangering one's moral integrity. Those who could no longer hold back the pressure of the concealed refutations were in the most difficult situation. Their 'background consciousness' burst open, exploding thereby the credulousness of 'false consciousness'. For example, in the first wave of the great trials of 1938 the manipulated evidence was accepted not only by the ones who were standing trial but also by the witnesses and other contemporary individuals whereas in the third wave, in the Tukhachevsky trial where the objective list of crimes of the false 'damning evidence' produced by the *Sicherheitsdienst* aggravated the charge, the doubt gathering in background consciousness exploded in an open moral resistance: the most important persons in the generals' trial such as Yakir, Putna and Tukhachevsky did not confess to their guilt and, moreover, Ordjonikidze even queried the very sense of the entire procedure. What at the beginning of the process had been only a suppressed doubt shut within the moral discipline of party loyalty now exploded as the staunch and moral defence of the issue. (All this of course did not yield immediate results, but this protest may have had something to do with the fact that after Stalin's death there were some politicians who could begin the moral rehabilitation of the movement right away in action. Without the explosion of these protests, the revelatory effect of background consciousness, this could not have happened.) They either tried to coexist with this strain or threw the conviction itself overboard, or made perhaps quixotic protests in the name of the idea. But I do not think I am far from the mark in assuming that one of the subjective sources of the power of the socialist system to regenerate is the way in which this background consciousness works. The 20th Congress started to lift the morally adopted or immorally produced controls which stifled these doubts but nevertheless preserved them and turned them into active forces again.

'Background consciousness' is, however, not an unequivocal phenomenon. Its operation did not prevent people from giving genuine credence to the manipulative measures and fictive formulae. Its function manifested itself precisely in turning conviction into belief or in preserving illusions. It represented only a possibility from which regeneration could later grow. It represented the possibility of self-examination, the sincere revelation of the past. But false consciousness lurks even in this revelation: 'I always knew it' is not only a moral self-deception (it is of course that, too). It is also the experience of the immediate eruption of this background consciousness. And the 'explosive' period of socialist development following the 20th Congress is also the consequence of the immediate eruption of background consciousness. Contrary to the false alternative of the Stalin era, a more recent false alternative often manifests itself in conformity with the logic of false consciousness: bourgeois democratic illusions or the bureaucratic reconciliation of 'everything has already changed'. Background consciousness is therefore not the refutation of existing false consciousness but only the unconscious notice of the corrections of reality. But this could take place both in false forms of consciousness and on the level of Marxist consciousness which seeks out and creates new alternatives.

'Background consciousness' is of course not limited only to the communist movement. It is a fundamental ideological symptom of our age. Even if we deny the theory of disideologization, there is in it nevertheless a sociologically true element, namely the resistance which has been developed against the ideological tricks of gross manipulations. The cataclysms of the 20th century have given rise to a psychosis of 'no man's land', a spontaneous apparatus of defence against all sorts of ideological effects and the expansion of the role of 'background consciousness'. In the illusions of the consumer society or in the gross manipulation of dogmatism today's man is incapable of consciously following through the chain of illusory relationships, so he has developed a spontaneous mistrust towards the news, the ideological facts and the exertions of influence. He has retained his credulousness, but combined it with the false consciousness of 'I must not believe it'. There is no doubt that modern manipulation already reckons in its strategy with this nonconformist attitude of defence, but in the industrial society this dormant mistrust at the bottom of one's consciousness is supported by such fundamental shocks as, for example, the Negro riots, the Vietnam war and the student movements which have dispelled the illusion of the 'ideal' way of life. The scepticism of background consciousness could thus become a false consciousness of the age. To the rebellious attitude and movements arising from this, Lukács correctly applies the term 'machine-breaking' (Luddism). They do not see the longer perspective or the possibilities for change. They are inspired only by the pathos of denial, whether they demonstrate against the police or formulate this 'no' in the sexual revolution, or strike an attitude of anarchism in the student movements. This 'machine-breaking' character follows also from the circumstance that what we had accumulated so far in 'background consciousness' as suppressed doubt, unease, self-deception etc. now became the false consciousness of public activity. And since it has formulated only a 'no' even as a 'background consciousness' it can get no farther than denial in its political platform either. The new varieties of false consciousness also encompass the technological rationality of our age.

This is where the mission of the renaissance of Marxism manifests itself in an endeavour to create from false consciousness the true self-knowledge and social consciousness of the age. It is a programme in the class struggles of the Western countries and a strategy in the alternative choices of the socialist countries. The role of Marxism, as an ideology of liberation, is to open up the impasses of false consciousness. It puts an end to the conflicts between the particular individual and artificial collectivity and reestablishes the balance of the totality. But this would be the subject of a further analysis.

CHAPTER 4

THE ILLUSORY FORMS OF EVERYDAY LIFE

> "I know many a game,
> for indeed reality whirls away
> and what remains is only illusion."
> (Attila József)

4.1. THE ILLUSORY FORMS OF EVERYDAY LIFE

In the foregoing we have taken account rather of illusions which present themselves 'for everybody' in the contexts of 'society as a whole' and mentioned at best their 'ramifications', i.e. the effects they have upon the horizon of everyday life. But in ordinary life the great blocks of illusions which pertain to society, economy and politics present themselves in combination with the ambiguities of the 'pseudo-concrete' world. These ambiguities come into being or die in the 'microcosm' of individual people and form in any case a nearly independent sphere. This medium is not unequivocally false, nor is it unequivocally real. It consists of small lies, pious frauds, hypocrisy, tactical manipulations, that is, of the hardly dissociable gestures of concealment and disguised forms of presentation. As a summary of the foregoing we now propose to reverse the course of our analysis and by proceeding upwards from below we shall analyze the forms of presentation of the formation of illusions and illusoriness. We shall start out from the ambiguities of the average man's sphere of life, a sphere which is near the average man by virtue of the simple reason that he lives in this medium. It is here that his cosmic disillusionment may occur, the discovery that 'everything is only illusion', that the utterance of reality (or of truth) is at least as impossible and improper as the insistence on open lies.

Everyday struggles move about in a medium whose inevitable criterion is this characteristic ambiguity, this 'quasi-reality' which has acquired an overtone of banality, where each gesture and word or all means of communication have been polished to assume several aspects. It is thus that this medium becomes the natural vehicle of this ambiguity. There is an extraordinarily wide range where this ambiguity, the 'microflora' of illusion, is prevalent. It is there, from the different rules of games and shifts of verbal emphasis in such games of benevolence, of love and power as are played with people, through the phenomena of intentional or accidental misunderstanding, to the formula of cynicism, hypocrisy, disillusionment and self-deception or the symptoms of defensive attitude or even pathological neurosis. Of course, we must not construe this medium as a mystical fate, not only because we also give and get some relationships in the exchange of the ambiguous communication and understanding, but also because it is a 'moveable category'. In certain situations, in the possession of certain traits of character, we can rise above this ambiguity or go under it under the pressure of either the circumstances or an ethical atmosphere. The greater proportion of everyday illusions have an ethical overtone. Within certain limits it also rests with the decision of a man whether he accepts that which readily presents itself for him or throws it overboard. The objectivity of illusions therefore means here, too, that they determine the widest bounds of play, ones which cannot be exceeded at all or, if they can, then only in exceptional situations, at the cost of heroism or perhaps not even then. And on the other hand, within these wide bounds it will already depend on the individual person's disposition and his ethical development to what extent he regards as 'ready' possibilities the illusory 'panels' which offer themselves and which also play into his hand through directions for use, from words through behaviour types to the 'responses' existing in the sphere of public or customary morality.

The illusory sphere of everyday life presents itself of course with a certain coefficient of inertia, often with a very high number of stresses. To step out of this sphere, to recognize the immorality of its stereotypes and indeed to see through its illusoriness, is extremely difficult, mostly uncomfortable and — perhaps — perilous. In this respect, our communication in everyday ambiguities has a considerable ethical overtone. The process of becoming an illusion is not an automatic one which is entirely independent of individual decisions. It is a process which is 'filtered' through the individual moral decisions and which also assumes an imperative fight with reality before each decision as if there were only a single alternative, the one that builds illusion. The narrowing down of everyday illusions or their extension — insofar as it is a sociological question — means at the same time the conflicts implied in the ethical conditions of these illusions. Whereas in the foregoing examination of illusions we repeatedly emphasized the objectivity of this medium — that its effects in the last analysis take the form of tendencies or objectification — in this everyday medium we shall perhaps be given the opportunity to show that this 'microflora' of social illusions, of which the illusory projections of greater relationships are also built, move in the range of illusions which have been 'made' and are being made. Thus, for example, play and playfulness, whether we take them as artistic categories or as social and behavioural ones, can have a liberating and heuristic effect and they can represent a cynical manipulation, a pattern of behaviour and self-rescue which conceal the harsh facts. On a higher plane of objectification the same applies to the objectification of art. As has been a commonly known fact since Lukács's *Aesthetics,* the work of art works with a 'created world', with the *'natura naturata'*: it works with a separate world which is, on the face of it, the same as ours, but this created illusoriness means only the technical framework of the way in which it is constructed. Socially this artistic illusion can be the revelation of the unknown conflicts of our life, it can have a paralyzing effect; it can have the peeling off of the stereotype 'husk' of our life and the cathartic representation of the naked reality. But it can also be an artistically engaging creative work, even initiating a new conceptual and artistic trend, subjecting only to a pseudo-revelation the illusory elements which function as reality, and thereby either involuntarily or in the by-path of 'the thought not thought over' (Attila József) recreate a secondary medium of illusions which has a seemingly 'oppositional' but in fact integrating effect. Dürrenmatt, in *The Visit of the Old Lady,* takes the former path, whereas Max Frisch, with his *Biography: A Game,* an analytical drama, decomposes the illusory structure and shows its lie, but in the end rearranges the structure of the original illusion from these decomposed elements. The final conclusion of his drama is that however much man would like to shape his life differently and whatever way he might decide at the turns of his past life, he would be unable to avoid the final outcome, the final decomposition. This is what we mean by the alternative characteristic of the illusory sphere of everyday — and artistic — life. We can proceed along the trace of benign, i.e., heuristic illusions where our destination is the reconstruction and the active acceptance of reality, or we can struggle on under difficulties as the prisoners of 'malignant' illusions or enjoy their comfort with a spontaneous ease. But the trouble is that, again as a result of the fundamental natural history of illusions, this alternative is concealed by immediacy, by the current of the spontaneous way of life. And here we are back where we started from, at the ultimately moral measure of moving in everyday illusions. The development of individual autonomy, the ability of rising to the occasion and the achievement of a historically possible record-breaking performance of man's 'species-being' constitute the yardstick and the watershed which make it possible to choose between the two different kinds of illusions.

First we propose to outline the natural history of the most innocent category of everyday illusions, namely the characteristics of playfulness and irony and their final ambiguity. Taking these as our starting-point we shall scan the increasingly more complicated system of hypocrisy, finer manipulation, misunderstanding, self-deception and, lastly, artistic illusion.

4.2. ILLUSIONS OF PLAYFULNESS, IRONY AND FREEDOM

Every game creates a particular sphere of illusion. It shuts its participants (those who accept the rules of the games) into an 'enchanted' world. It separates them from the real world while delivering them from its cares, and fills them with new worries which promise some distraction. Their joy is fuelled by the consciousness of this separation and at the same time by its illusoriness which has momentarily been forgotten. When we play we are not only aware of the 'unreality' of this medium but consciously, although we try to hide it even from ourselves, strive for the preservation of this 'falseness' and for the well-oiled functioning of its mechanism for forgetting. This partly conscious and partly unconscious self-deception forms the basis of the self-enjoyment of play and its 'joy of reality'.

We cannot aim at elaborating the central plan of the entire theory of games, and this is indeed not even our purpose. We would only like to explore the clue with regard to its special power to create illusions and, in this respect, to find the issues which convey the marks of conformity to and difference from the hitherto analyzed components which form illusions. We therefore cannot enlarge upon criticizing the various theories of games which have been ever on the increase since the beginning of the 20th century, criticizing the papers which have analyzed the phenomenon of games from the point of view of anthropology, child psychology, psychoanalysis or sociology. What I have in mind here is the analysis of competitive sports, games of chance as well as the various forms of 'passive playfulness', such as e.g. the spectators' 'game'. They are described, as well as superficially rather than thoroughly analyzed, in e.g. Roger Caillois's famous work: *Les jeux et les hommes* [Paris 1958]. We had better focus our attention on phenomena where playfulness presents itself as a social attitude, as a disposition, namely where the object of the game is not an alienated thing (a card, a football, a team) but 'another' man in society or perhaps even ourselves. It is only with a view to give an idea of the subject that we refer to the conduct of Thomas Mann's Felix Krull as well as to the analysis in which Lukács calls our attention to the typically modern way of life that this character represents. [Cf. Gy. Lukács: 'A játékosság és ami mögötte van' (Playfulness and what is behind it), in: *Világirodalom* (World literature), Budapest 1969, Vol. II, pp. 192—224.] It is against the totalization of illusions that Krull puts on the 'space suit' of ambiguity which protects him and at the same time also helps him to enjoy the usufruct of the 'world situation', and by this means he turns the instruments of the false reality of the modern bourgeoisie against the same world with maximum results. His tricks and lies are accepted more joyfully and with greater trust than his contingent, accidentally occurring honest manifestations. Krull plays because the world plays with man, and since he has found out the secret stratagem and totalization of manipulation, he can be a winner.

The philosophical and everyday attraction of play enters 20th-century philosophy, literature and the order of behaviour types as the intensification of a 150-year-old tendency. This attraction of historical philosophy has lived in the thought of modern times mainly since the trough following the Enlightenment and the French Revolution. The notion that play will be the way of life of mankind's liberation is, as it were, a steady outgrowth of the utopian and artistic dreams woven about a new society. Schiller already proclaims this principle in his *Letters on Aesthetic Education* [1793—1795]. This idea will also be one of the central categories of Fourier's utopian social system and historical philosophy, to say nothing of the significant character-forming role which it plays in Mozart's world picture.

But of course even if we juxtapose the conceptions of these three playfulnesses, we can speak only about the final note of three radically different historical philosophies which have been given a common designation. Playfulness with Schiller is little short of a fleeting thought in the *Letters on Aesthetic Education* which does indeed chime in with the essay he wrote on 'grace' [1793], but still remains isolated in the entirety of his works as a brilliant germ of idea. The possibility of realizing this idea at most flashes out again — in a more resigned form — in *William Tell,* a utopia for the revitalization of the popular form of life. With Fourier, as is shown in *Le nouveau monde amoureux* [1817—1822], which he left in manuscript, this is an essential element of a total conception of historical philosophy. His dream is to reproduce for adults the enjoyment which children derive from regarding their amusements as work. He treats it as one of the dominant motives of a transformation in terms of utopian social criticism. But he already foresees some of the oncoming results of the technological revolution and so this daydream does not prevent him from treating, collaterally with this playfulness, the nearly automatic dictatorship of the modern forms of work with the same kind of utopian enthusiasm which can be understood as critical overtones and idealized balm alike. Mozart on the other hand never formulates playfulness on a theoretical plane. He only senses that this element is also present in the air in the great convulsion of the period, in the inconsistencies of the ideal of freedom, and it is through his contacts with freemasonry that he champions the cause of a musically constructed and at the same time increasingly gloomy ideal of play, ending on a note of tragedy (*The Magic Flute,* and his late symphonies — G minor, E major). Be that as it may, thought is the offspring of the self-redeeming pathos of the age, representing a certain vigour of liberation even in this utopian and contradictory form. Without contact with this conceptual matter, Diderot's or even Rousseau's anthropocentric system and concept of naiveté would have been inconceivable. (This 'influence' is of course not derived from Fourier who wrote his major works well after the Napoleonic period.)

In these conceptions play represents a protest against the distorting effect which the capitalist division of labour and the capitalist type of work have on people. It is a paraphrase of the liberation of mankind. It is undeniable that playfulness, as an intellectual behavioural faculty, gives us a certain relative surplus of our disposition. It provides us with a distance against the events and the people that surround us and, therefore, raises us from spontaneity and at the same time sinks us into the experience of a different — in a given case more normal — spontaneity. Playfulness of course also exists in a bureaucratized form, where it is the alienated 'distance' of the trifling with people, with data and with documents, where the source of joy is an inhuman self-indulgence. This relative superelevatedness therefore means only that we succeed in rising from the sphere of immediacy and that we can, at least within the bounds of possibility, examine and handle our object and our connection with it from several angles. This superelevatedness is relative since by its nature play can create only a virtually free world which only a new effort can transform into a real one. But, as far as possible, this real liberatedness can still have an area of preparation, an experimental medium. And of course it is indeed this possibility which gives it its ambivalence. In other words, there also exists an alienated type of play regarding which we can no longer speak about either a moral or an intellectual superiority, where the reign of immediacy may obtain an absolute domination in a roundabout way, in the disguise of virtual freedom. A precise description of this phenomenon (precise because it is also alienated in the writer's capacity) is provided by one of Samuel Beckett's radio plays, *Krapp's Last Tape,* where the single character who is vegetating on the verge of senility amuses himself by playing back the events of his life which he recorded some thirty to forty years ago, and by doing so he relives his former life. Unalterability is in the same way manifest here as the destructive effect of the bureaucratically alienated play. Krapp derives his pleasure from cataloguing his past experiences, putting them in a box, selecting them later, adding his present remarks and then replacing them in their mute solitude. The system is badly organized because he becomes absent-minded in his old age, but the order itself that he attends to is both a game and a negative self-indulgence. It is the already joyful or at least contemplative observation of his own ignominy, of his one-time tragic mistakes. He enjoys the impressions of the odd mixture of the could-have-been-differently and the it-happened-only-in-that-way. (On the occasion of an amorous outing Krapp, as a young man, presumably killed a girl or drove her to her death, a girl whom he loved indeed but with whom he did not dare or did not want to assume the obligations which love implies. And ever since then his life has gone on the wrong track; his living putrefaction is, as it were, not only the 'punishment' for this disastrous event, but also the motive of this bureaucratic replay, his self-justification.)

Even if we must establish that play has some dangerous effects which might tempt us into unreality and which might make it easy for us to assume a bureaucratic attitude, we cannot deny its virtues. One of the dominant notes of Attila József's poetry is the state of childhood and within this the longing and demand for the return of the penchant for play, which is not only the result of the proletarian's destiny or the grief felt for having been deprived of the state of childhood but also the real and consequently no longer utopian element of a positive human ideal. It is due to play that he can avoid the 'powers that ruin', the otherwise inescapable track of alienation, and find the first active possibilities of becoming a consummate man. What with the communist Attila József assumed the form of a distinct and deliberate strategy, a yearning which created poems and imposed a form on conduct, was only the crowning of one of the tendencies of 20th-century lyric poetry. Already Dadaism and Surrealism had begun to playfully decompose both poetic and social forms and the ways of life (from Apollinaire to Breton and the lyric poetry of Aragon's young days), most of the time only suspecting that there was more to this playfulness than the possibility of self-enjoyment and shaping of human character by means of destruction. What with them had been only a germ became with Attila József a poetic programme. And the odd thing about it is that Aragon and Éluard go beyond this stage and at the same time discard these playful overtones (retaining them at best among the elements of their love poetry) and cannot preserve them for an 'objective' lyric mode as Attila József did.

The situation is more complicated in the anarchistic or infantile ideology of modern student movements where, as the playful and liberated destruction of standards and corruptive material wealth of the establishment, playfulness comes to the fore, whether in the sense of attraction to violence or of absolute denial, or in the name of an utopian and vague conception of the future. For lack of a positive ideological centre of coordination, regeneration is accompanied here by a spiritual and moral regression. It is on the other hand probable that this playful element takes a prominent part in the development of socialist ownership even if it is not in this 'playful' way that it is put into practice in the reorganization of society. A case in point is the fate of cars 'taken over on socialist trust' or of land farmed by the brigades 'as their own'. Experience shows that a much greater efficiency could be got out of people when the driver or the brigade could functionally, that is to say playfully, believe that the machines, the cars or the land were their own, and that the protection and the durability of their substance or even the external appearance of these things scored much higher 'points' than in the realistic forms of 'impersonal public ownership'. People tended and even decorated these means of production 'as their own', and it was due to this quasi-playfulness that they made at length the 'spiritual' expropriation of the publicly owned means of production acceptable and a matter of course for both themselves and society. Of course, here we are talking no more about play in the literal sense, still the illusion-creating elements of this sphere (which give rise to an illusion with a positive tendency) are present and have a playful effect, but only because the total structure of society afforded the possibility of such a tendency.

Yet the modern bourgeois way of life gives less and less scope to the field of force of the positive possibilities of play and uses more and more its virtual or indeed destructive efforts to offer seductive possibilities for role-playing. From this point of view we may perhaps regard the modern development of French films as documentary evidence. Love, putting oneself to the test, or even innocent games carried on with others, are shown in these protest films as the ambivalent products of the 'consumer society'. It is as if the liberating potencies had been completely wiped away. To illustrate the point, there are the first important products of the 'new wave', such as *A bout de souffle, Jules et Jim* or the satirical and ironical charges of Jacques Tati in e.g. *Playtime.* The play in Antonioni's *Blowup* is a distinct symbol of travestying a society which is built on illusions. When the photographer who is the principal character in the film picks up the non-existent ball and as a partner joins the absurd game of the carnival's masqueraders, it means his self-surrender, i.e. the surrender of his individuality and with it his human possibilities which glimmer for a moment. Here play has once again lost its meaning of self-education and self-fulfillment. It is a self-indulgence in which one can enjoy only one's destruction. But even in this murderous illusory sphere of alienation we cannot assert that this playfulness (meaning an artistic manner of description this time) lacks all elements of protest or those elements which experiment towards real freedom. Think of the nostalgic, fairy-tale representation of play in the renowned film entitled *The Red Balloon,* where play manifests itself as the bitter hope of a better world. Or there are the factual novels by the American writers who use the genre of exposé to pass social criticism on the playful way of life [Capote: *In Cold Blood;* Updike: *Rabbit, Run*]. In these books the mass murderers, under the influence of an aberrant social psychosis and not the least because of the pedagogy of the senseless war, kill — as one murderer admitted — 'just for fun', 'naively and playfully'. As the instrument of negative, black irony, playfulness has turned here into a murderous criticism of society.

It was Marx who pointed out the ontological causes of this ambiguity of play[1] and categorically asserted that with a view to the perspective of human liberation, the utopia of a 'playful society' could not qualify as a fundamental solution. The 'realm of freedom' can never be realized on a playful basis. Man was made man by work, and society can exist and can mould itself in work only as a member of the 'realm of necessity'. Play

[1] In the *Grundrisse* he writes that the increase of leisure time is the abstract condition of the self-development of the individual and that with it play, too, can be taken into consideration 'as a terrain of experiment'. But work can never be play as Fourier would like it to be, whose undying merit is, however, that he stated the final judgment not of the conditions of distribution but of the decomposition of the mode of production and its transition to a superior form. Leisure time which includes both time spent on entertainment and time appropriate for some superior activity of course transforms its owner into a different subject who will step back into immediate production as an altered individual. Leisure time rendered in this sense also requires discipline if we regard it from the point of view of the forming man in whose mind there is the accumulated knowledge of society, as it is the careful study, experimental science, i.e., a science which creates materially and objectifies itself, when regarding it from the point of view of the existing man.

and with it the 'realm of freedom' can take part only in the formation of leisure time. His theory of play as well as the liberating effects which as elements really dwell in play are functions of the quality of the prevailing structure of work just as the character-forming or deforming power of leisure is the function of the alienated or anti-alienated quality of worktime. Marx therefore does not deny that play has a liberating role which is there as an element. On the contrary, he emphasizes a different role of play which transforms a man and which objectifies him in an intrinsic and social disposition, but only as a function of real social practice, i.e., work. The two spheres, leisure time (and play in it) and work, as a total activity in which they form each other, become the educators of mankind. The above-described ambiguity of play follows from this dual restriction. It follows, on the one hand, from the positive liberating possibilities of its self-movement and, on the other hand, from the consequential effects of the frameworks which affect it from the social conditions of the 'realm of necessity', i.e. of work, and which curb the realistic endeavours for self-realization of man who wants to be or has already become something else in play.

Let us examine now the special feature of the sphere of illusions which play can create. For one thing, play is an objectification itself just as work is. It manipulates either with the objects of the real world or with people, but accepts them and handles them in a virtual role and, as such, as the vehicles of these roles, objectifies them as real elements in these roles. It changes them into whatever categories they were assigned to in the course of the game. Hence it has a virtual objectification. But the acceptance of the objects used in the nature of a role in conformity with the virtual requirements and contrary to reality is objectively binding on the players (at least while they are playing) in the same way as are in real action the instruments and the rules of instruction pertaining to their handling. What was originally the product of imagination is now mutually accepted by the players, for these objects convey this role and this form of presentation for them. It seems to be the objective characteristic of things because this is the way in which play objectifies them. Here there manifests itself the element of play as freedom. It suffices to decide, i.e. teleologically suppose, that this chair is now a spaceship, that it is a friend of mine, my lover's passion or character, the 'object' of my escape or of my need for love, and I can immediately regard it as such, provided that there are at least the two of us partaking in this relation. It is these objective media which convey mutual making. Their objectivity is formed in the field of force of play which creates thereby a 'different' world.

Play does not simply 'project' an alien system of pictures on to the objects of reality but works in fact with a dual objectification. There are two planes in play which simultaneously interlace with and separate from each other. One is that which the hand and immediate practice do and the other with which teleological imagination and practice play. It is this contradiction which implies the magic and the wonder of play. And indeed, even the smallest child experiences this contradiction simultaneously, and it is due to that that the adult, too, finds enjoyment — even rapture — in play. What do we mean by this contradiction? For the sake of simplicity, we once again refer to the simplest of children's games, the making of sand cakes. The two planes, i.e. play and real activity, are interlaced as a matter of course, but viewing it from the outside they are also distinctly separated from each other. They play with sand but they still see it as a cake, and while they handle it as a cake they still take good care not to lose sight of the fact that it is of the nature of sand, namely wet sand is needed for making the cake, some 'flour' has to be laid on the forms etc. Notwithstanding that on the plane of play they handle and regard it as food, practice cannot delude itself. Play will be successful only if they observe the requirements of the original materiality. And likewise, everybody will 'consume' the cake once it has been made, but none of the children will actually eat it. They see it and make it as a real fine pastry and they also see it in the shape of sand. It is only the teleological directedness for which the cake-like resemblance is more important and 'more visible'. So teleological objectivity proceeds in a medium that is different from the one in which activity factually handles the object. Both a distinction and an identity must continuously be established between the two. It is this dialectical tension which inspires the magic of illusion. The point under discussion is therefore not that there is a difference between the way they make and the way they see the object of play. This would be too simple a differentiation. What is characteristic of this form is rather that one must simultaneously act and view in two planes. The object being formed in one's hands is 'amphibian' (both sand and pastry) not only for practice but also for observation. Should reality remain only in handling and playful objectivity only in observation, play would lose its characteristic zest: the surveying and voluntary 'forgetting' of unreality, the self-deception of an illusion which created itself.

But teleological objectification not only forgets the original objective characteristic (the stick which everybody accepts 'as a horse' is not perceptible, the piece of wood is a battleship, the chair a spaceship etc.), but it is in fact the original 'different-in-kind' objectivity which will be the condition of the separate world of play and of its objectivity. Failing this, there can be no partnership, no objective and comprehensive order of common activity. Real objectivity will be the vehicle of the playfulness because it mediates the mutual relation of the players. Someone playing alone can create an illusory sphere even without such a 'supporting surface', but this cannot be regarded as true play since there is no difference between the conceptual action of imagination and the activity which works with genuine objects. The essence of play is that we can sense that its world is objective towards us and that we depend on its rules and its objective conditions even when we are conscious of our self-deception.

Play is from the outset a social relation. It occurs only when this picture of imagination has already become detached from me and assumed an objective configuration which is independent towards me, too. That is why it is necessary that teleological objectification take an objective shape, that something in the form of either tangible and 'makeable' objects or some rules convey the object which is being playfully handled. Everybody makes sand cakes in like manner. The objectification of common imagination is conveyed by the order of common and similar making.

The task of this object is to create an analogy, to evoke the fictive object. Since in one respect it is 'like' what it figures as, in the analogical mode of perception and action it will already function as that does. Teleological imagination immediately projects onto it the characteristics of the other object. This is how illusion is created: I see it as 'that' because it manifests itself as that and everybody else reflects it to me 'as that'. As we can see, here the category of 'object-making' (cf. Chapter II, Section 3) has returned in a fictive, virtual configuration. What was a part of the objective structure of social practice there, here becomes a means of creating a world outside reality. 'Object-making' is a law which is common to both.

The situation is similar as regards the more complex forms of play, where a conceptual objectification of the rules takes the place of the immediate material. We must 'observe' these rules and we must act in conformity with them. But play itself is not confined to rules which, like the 'material' of play mentioned above, constitute only a controlling medium of support. (It is a different matter that the violation of the rules is also a part of a game, and this violation may be a game itself because it is here that it becomes a true experience and that we can enjoy the fact that we are dealing with unreality.)

The space which encloses the players is objectified in the objects and in the rules of the game, although originally it is the player's teleological notions which form the 'separate world' and the special nature of this space. It is the objects themselves which require to be handled 'as what they are', but it is also evident from the partners' responses. Each object will figure in this sphere 'as something else' and be real in this function without hiding its original structure once and for all.

The Enlightenment could regard play as a factor in the liberation of mankind because in play we can truly free ourselves from the objective necessity of the material world and we are subject only to the perspicuous and satisfiable requirements of a necessity (rules of the game, the conventions) which we create ourselves. Everything is repeatable and reversible. If I want to I can call 'it does not count', and if the objects do not obey us we can pretend that they do. On a virtual plane play brings purpose and realization in close proximity to each other, though in real life they are farther away. Play is a purposeful activity which can realize its desires much sooner in quasi-reality than in everyday life. The attractiveness of play is lent by the approaching realization, the attainability of a purpose, as well as by the experience of having, as possibilities, the decisive instruments of success in my hands. But not only this; its illusoriness has a structure which is easy to survey at any time. It does not work with a fossilized illusory aspect the reality of which cannot even be suspected any more, but it creates an 'as if world' in using which we enjoy both its separation from reality and the illusion of this separation, the magic of illusoriness and the joy of revelation. This element also has a suggestive influence on a man who is surrounded by the baffling illusions of the modern way of life, for indeed he can move about in a world which he can survey and in which the tractable illusions which have been laid open are all in his power. This immediate reversibility is the special feature of play. The real world and the playful one are both present in illusions which glimmer through and exist by means of one another. Reality is merely a role of the playful world just as playful configurations are of reality.

The bourgeois theories of play, insofar as they have been written with a view to giving an overall picture of a philosophy of history, proceed from the circumstance that 'everything is different' in play since it transforms the world with the help of 'as if'. These theories have two ramifications. We have already referred to one which sets out from the illusions of the Enlightenment. Its essential character is a protest against the distorting effects of the division of labour, a particular confrontation with reality.

This illusion runs from Schiller, Fourier and Mozart to the artistic practice of irony and playfulness with claims to realism (Daumier, Shaw, Thomas Mann, Dürrenmatt). The thought content of the other line takes an opposite view. It is a theory which is based on a reconciliation with reality, on a playful and superior resignation, a theory of 'sapient smile'. It starts in German Romanticism (Friedrich Schlegel, Tieck) and has lasted to date. By means of Kierkegaard's intermediation it runs through Huizinga to the theory of roles of existentialism, and as an artistic look and principle of form to the reduction of play to the absurd or to the one-sided expansion of the ironical outlook and mode of life. The *action gratuite* which Gide invented and depicted had its genesis in a fetishized and at the same time cynical conception of freedom. Camus in *The Outsider* describes Mersault's murder, a nearly self-liberating playful apathy of a murder, as a form of action which is an extreme case of irony and obscurity. But the sadistic and hedonistic experiment carried out on a mere slip of a girl in *Lolita,* a semi-pornographic novel by Nabokov, is nothing more than a piece of playful and cynical self-indulgence. The characters in Kerouac's *On the Road,* who dwell in a frenzy of liquor, reckless driving, marijuana and beat music, live almost exclusively in the desperate stupor of this game of life which has hardened into a game of escaping reality and which has been deprived of its playfulness.

We have mentioned that playfulness has two possible directions in which it can accomplish itself. This dichotomy can also be found in irony, namely as regards its role in life and in artistic representation. Undoubtedly, just as play has a characteristic feature which can create a distance so irony can also represent an intellectual mien which relaxes the alienated customs and the obsolete standards. Socratic irony and the early Enlightenment of Montaigne represented the first steps concerning the liberation and even the moral accomplishment of the human spirit, as a transition to creative scepticism and then to rational denial. But this is only one of the possibilities. Irony, deprived of its perspectives, can withdraw into itself, and it can also be a hoity-toity aloofness not only towards the 'world' but also to ourselves, i.e., the pleasure of the outsider's existence at the cost of an alienation which it wants to deny. Here irony offers an illusory superelevatedness and a virtual freedom, barring the possibility of a next step and thus getting in the way of the gestures of real liberation. Let us turn our attention to the most authentic theory, to Sartre's concept of ironic playfulness. The worldliness, the attainability of man's conception of freedom, is ensured by the fact that he considers ironically all his obligatory vital functions which have been socially laid down for him, and he takes them on only as playful roles, handling them as a disguise which he has assumed, as it were, by an incognito, so that in the meantime he can preserve the autonomy of his human, moral core. Man only 'plays' social standings. He pretends that he has formed those himself and thereby suspended the shocks of their objective commands so that he can regard his playful independence as real. Sartre demonstrates this dialectic with the following example: the waiter is not immediately identical with his duty, for indeed as soon as he consciously performs his work, i.e. treats his professional tasks as a role, he is already play-acting. "He is playing at being a waiter in a café. ... The game is a kind of marking out and investigation. ... But if I represent myself as him [as the waiter — M. A.], I am not he; I am separated from him as the object from the subject, separated by nothing, but this nothing isolates me from him. I can not be he, I can only play at being him; that is, imagine to myself that I am he." [J.-P. Sartre: *Being and Nothingness, Op. cit.,* pp. 102—103.] Play can dissolve the sternness of reality, it can perhaps even annihilate the existing order of the world.

He does not raise the question of whether it is really so easy to give up the system of duties within which the individual can play. The initial stage cannot be problematical for the individual because people accept the objective roles of reality as a matter of course and they play only within these roles, never impugning their frameworks. They make their conditions of existence unreal only on the subjective plane. They can imagine that just because they play with these conditions they do not exert their influence with the force of necessity. The playing man — as Sartre says — is also free, for indeed it is not the world but he himself who dictates his actions. But the trouble is that this ideal of freedom does not repudiate the factual obligations which social roles imply but only phantasmagorically relaxes the implacability of the obligations of life. Objectively he is united with the everyday requirements by bonds just as indissoluble as if he did not even play. Irony and playfulness help him at best to endure the agonies by consoling him with the illusion of freedom. The waiter who plays at being a waiter can more easily adapt to his profession because he can forget that he is defenceless and because he can imagine that he is something more. Here irony is the artificial reconstruction of human dignity and freedom as well as the instrument of their actual loss.

This type of irony, insofar as it offers the illusion of freedom, is the solace of everyday drabness. What with Sartre was a protest against the modern phenomenon of alienation was in fact the further development and the generalization of Romantic irony. It was Romanticism which discovered the subjectivistic ruse that so far as man seemingly accepted the order of reality he could even query its trueness, its sense and its authenticity under the cover of this gesture. Ironic assent has always been of course a covert form of denial, whereas assent has represented the destruction of a thought structure. From Socrates to Erasmus, classical irony resorted to this conceptual attitude to ridicule the standards and the formulae which could not be queried overtly, without the necessity to lay down the doubts and the objections as principles. It merely asked questions or strongly emphasized the acceptance of the formulae which were suspect, at least to the extent that their absurdity would be obvious. Socrates destroyed the arguments of his opponents by entering into the spirit of the game and then letting his opponents get entangled in the inconsistencies of their own arguments and seek the genuine solutions themselves. As compared to this classical irony, German Romanticism contributes something new. Romantic irony serves to induce inaction, a playfulness which exists only as a mode of perception. It does not want to become an intellectual opposition which would give rise to practical resistance. It is an endeavour to create the mode of life of an inward independence in which there is no need to launch an open attack against the established order, while at the same time one can still smile contemptuously at the foolish world from the zenith of an illusory superiority. The purpose is to show approval both in practice and also by means of the approving actions, while inwardly, as regards subjective conviction, ridiculing. This attitude is therefore precisely the opposite of classical irony where it is an intellectual attitude which is the germ of an active intervention at some later time. In Romanticism irony is a practical life-principle, the release and the consequence of which form the subjective stability, an inner independence, that is to say the illusion of freedom. What in classical irony turns to action here escapes from an action which accepts illusions, to an illusory freedom which is incapable of action. Hegel has a sarcastic characterization of this attitude and of this artistic frame of mind: "I live as artist when all my action and expression whatever, in so far as it has to do with a content, is for myself on the plane of mere semblance ... So I am not truly serious either about this content or ... about its expression and realisation." [Hegel: *The Philosophy of Fine Arts,* transl. F. P. B. Osmaston, London 1920, Vol. I,

p. 89.] "No doubt for others my self-revealment, in which I appear to them, may be taken seriously, in as much as they interpret me as though in reality I was in earnest about the business; but therein they are deluded, poor, borne creatures, without the faculty of the power to comprehend and attain to the height of my arguments." [*Ibid.,* p. 90.] Incidentally he also points out how this attitude is associated with Fichtean subjectivism and shows the consequences to which it leads, namely to the draining of individuality and to the 'looseness' of the Romantic frame of mind.

Romantic irony manifested itself originally only as an artistic principle of form, i.e. as an aesthetic principle. Kierkegaard is the first who sees a general social attitude in it, i.e. a certain type of life-style. Kierkegaard launches the modern idea that life must be played because it is only thus that a man can be free. In his first philosophical study [*Über den Begriff der Ironie,* 1841] he proceeds from the idea that playfulness liberates man from the roles to which he has been assigned by the division of labour, and that, though it is only negative freedom, it may nevertheless mean man's only positive possibility for self-realization. The ironic attitude, says Kierkegaard, either identifies itself with the absurdity which it assails or opposes it, but the way it does this is that it is always conscious of its illusory nature and it always does indeed enjoy its unreality. Under the circumstances the possibility for initiatives is within its capacity, and its activity is no longer determined by the earlier conditions. In this imaginary form reality loses its validity and man becomes free. He destroys reality through concrete reality itself. [Cf. Kierkegaard: *Über den Begriff der Ironie mit ständiger Rücksicht auf Sokrates,* München—Berlin 1929, pp. 207—211.] Irony plays identity and opposition as one and the same and thereby renders the world unreal. It is not necessary to shatter the system of these roles which ruins both the genuine man and relations but it suffices to play them ironically and they will lose their validity at once. The initiative does indeed get immediately into the player's hands: the waiter in fact chooses himself how to play this role, but it is only the illusion of immediacy, i.e. self-deception.

Kierkegaard does indeed believe in the quasi-revolutionary character of this attitude which, as he says, changes the existent into illusion and, consequently, causes its perdition. [*Ibid.,* p. 227.] Yet at the same time he sees clearly that this 'perdition' and this release are only subjectively valid and thus only negative.[2] Sartre loses sight of this dialectic in which the consciousness of illusoriness could still be present. He already considers the notions of subjectivity real and fails to notice that it is the objective structure of society that determines the system of duties which can be 'seen differently' and dissolved into the forms of 'as if' but which cannot be hidden through intuitive and practical retouching. The waiter will be a waiter even if he plays his part adroitly and consequently happily.

But, of course, however critical we are in our manner of approach to the Romantic and modern conceptions of irony and play, the truth-content of these vindications depends on the proportions. That is why we must emphasize once again that irony, as an artistic attitude or even as everyday wisdom, in its element, in one of the directions it may take, represents the germ of a real social-critical or indeed liberating type of action. And though this question has already come up several times in the preceding pages, we must nevertheless refer to the important achievements of modern literature, from the best writings of Thomas Mann and Dürrenmatt to some of the top performances of American literature where ironic vision ranges over a wide spectrum; but it is exactly this attitude which helps to penetrate the fog cover of social illusions. And on the other hand, there exists today an annoying stereotype of feigned literature, namely the use of irony as a style, which stems from the manipulative 'integration' of this method. The external criteria of the ironic mode of perception, its attractiveness in such matters as keeping its distance, its propensity for humour and its critical acumen in implied approval, have continued to exist, but only as trappings, as a writing technique under the guise of an artistic world of conformity and a literary attitude which claims to protest at the outside only against superficial phenomena. In other words, the convention has grown up of making a conformist and fashionable style of irony and with it an intensification of the tendency of this attitude and mode of perception to go off the rails. (In most cases the absurd dramas represent an inner artistic despairing of a radical attitude of critical acumen which can dig down to the human roots of society, though 'black humour', lyric irony as a technical medium, makes a much stronger impression than the classical products in

[2] Irony is therefore the extremely easy game which we play with nothing, and we are not put out by this Nothing, but on the contrary, it is here that for the first time we can hold our head up straight and high. Irony denies everything and in this respect it is absolute, for that in whose name it denies is a superior being which in fact does not exist. It is irony that establishes 'Nothing'. [Kierkegaard: *Op. cit.*, p. 218.]

their ironizing compositions.) Just consider Ionesco's half critical and half conformist writings. On reading his *Diary* we can right away detect the real value and the cynical nature of this attitude. The same can be said (although presumably accompanied by a purer literary purpose) of the line of decadent achievements, such as the play-writing of Max Frisch, the half bestseller and half literary work of Joseph Heller, the caustic humour of Harold Pinter, or the satirizing protest pieces by Kopit which are after all reduced to nil in their material as regards powerful emotions. Irony has already lost its charge in these writings. But of course irony represents even today an 'open' artistic mode of perception which when possessed by an important writer can still yield an outstanding result. The world-wide success of the play, *Marat/Sade,* by Peter Weiss came from a peculiarly interpreted irony which, in a severely critical and philosophical manner, further developed irony itself. Such top literary achievements are, however, becoming ever rarer.

Romantic irony certainly did not arise as a theory of freedom. It was rather the bitter day of reckoning of that theory and the quasi-protest performance of the acquiescence in the unreality of this slogan. Modern theories of games represent a further development of this tendency, with the difference that the awareness of the illusoriness of freedom is not so clear as it still was in Kierkegaard's writings. (He was still alive to the fact that the ideal of irony as freedom was "a height which in fact does not even exist". [Kierkegaard: *Op. cit.,* p. 218.] Such a sense of proportion became at best clouded by the middle of the 20th century.) At the same time, as a practical life-principle, the type of everydayness of modern life belied in the ambivalent form of brutality and false sincerity that which this theory of irony had propagated for a century and a half. What I have particularly in mind here is not the moral code implicit in the sanctioning of brute force, the cynical shrug of the war criminals or the theoretical cynicism of 'justified means', though this tendency, too, was released by the 20th century from the esoteric secrecy and made not only common knowledge but also, as it were, a customary attitude. The matter to which I refer here is that which has become a life-principle in the everyday version of this attitude in the most different kinds of nuances, from the routine cynicism of the sergeant obsessed with the spell of 'minor power' to the self-surrender and amorality of 'it cannot be done otherwise anyhow', to the avowal of irresponsibility as an open maxim. For indeed, the fact is that the 20th century not only continued to build the dreams which Schiller and Kierkegaard had had about the realization of the life-principle of 'aesthetics', of the 'inner freedom' which could be attained through playfulness, but it simultaneously brought about the everyday practice which proceeded to realize these theories profanely but in such a way that it also denied them. It became clear that if the choice was possible only among the illusions of freedom then the open and brutal abandonment of the inconvenient because unrealistic moral commands, the customary norms or the shackles of legality, might be more real or at least more enjoyable. Everyday cynicism initiated this profane 'revolution' and with it a quasi-real release in which a man could move about in a more unrestrained way, that is, he could enjoy the illusion of freedom on the level of practical reality. Playfulness, in the sense of Schiller's 'grace' or its modernization by Sartre, could only promise man that 'behind' the unconditional acceptance of social obligations he could do whatever he wanted to and that, provided he had sufficient perseverance, he could perhaps maintain his moral autonomy until time and opportunity would present themselves for its realization. This promise, however, only seemed to be temporization, for indeed cynicism offers immediate gratification. By assuming irresponsibility and abandoning the standards one can at once rid oneself from most of the constraints and — to tell the truth — there are more tempting elements of freedom in this gesture than in the different versions of fictitious independence. And furthermore, playfulness even lasts: one can use cynicism to play with people, to play upon their emotions or good faith, or indeed upon their convictions, in the same way as it was possible to play with respect to ourselves on the scale of fictive possibilities in the 'one-person playing-field', in the playfulness conceived by Kierkegaard and Sartre. This everyday and routine-like cynicism evidently could not take on such a many-coloured, theoretically perfected character as irony and playfulness could. By its nature it was regarded as a baser attitude, whereas irony, by virtue of itself, always rated as a 'more genteel' and supramundane type of attitude. Cynicism was a matter of practice rather than of theory. Its spell and its influence are to be sought in the free and easy immorality which makes life bearable and in the illusory emancipation of oneself from ethical standards. But this operation can only be practised (and even then only in a smaller circle at the outside), and propagated as a proverbial profane wisdom (so the key watchword, 'poor money — poor football', has gained a wide currency also outside the domain of sports). But due to hypocrisy or the essential need of society for an ideal, its theoretical summary and proclamation are nearly out of the question: it is bad manners to wash one's dirty linen in public. On the other hand, this is precisely why it is difficult to find the trace of the principles and the historical development of this attitude. We can rely only on conjectures, on some literary abstracts and on period documents which were formulated by conscious or unconscious candour and condensed in memoirs. Regarding the barbaric cynicism of fascism we have in mind the surviving diaries of Rauschning, Hoess (commandant of the Auschwitz extermination camp) and Goebbels, or the cynical gestures arising in the manoeuvres of the allies, Churchill's celebrated memoirs and also — as a literary product — Rolf Hochhuth's *Soldaten* [1968], a partially successful documentary piece which refers with moralizing unsureness to this phenomenon. The memoirs of Ehrenburg, Simonov and Zhukov allow one to get an inside view of the factual material of cynicism resulting from the distortion of socialist development, a cynicism which on a world-historical plane has a different orientation and a different system of ethical and ideological conflicts. It is of course only on the level of facts that these writings give an account of the different types of cynicism, mostly of selected facts or those which have been filtered through the different modes of perception or the confusion of information. The murder of Kirov, unsolved to this very day, and the suicide of Ordjonikidze, as well as the moral and philosophical implications[3] behind the gesture of the great political leader who paid his respects and stood in funeral guard belong, of course, as a question-mark not only of morality but also of sociology, to the issues of an age which have yet to be brought to light.

4.3. AMBIGUITY, MISUNDERSTANDING AND LYING — THE IMPERCEPTIBLE 'QUASI-REALITY' OF EVERYDAY COMMUNICATION

'All Cretans are liars', said the Cretan man. So runs the famous logical paradox, the *'Krokodilschluss'* ('crocodile syllogism'). But the first premiss is not true since there is no situation, group or country where it could be ascertained with such logical clearness that everyone lies. It is evident that there are also people who remain silent or who, out of naiveté or indeed for the sake of honesty, will tell the truth. The problem of 'truth' *('Echtheit')* starts at this undifferentiated generalization. This

[3] With regard to the different types of cynicism and its philosophical implications as well as its present-day forms of combination, for further details see my study *A cinizmus dicsérete* [In praise of cynicism], Budapest 1967.

premiss arises not only in the logical games of Greek origin (only as a game then) but also in modern sociology. We have already quoted Adorno's bitter remark according to which truth and lie are indistinguishable in present-day culture because truth has lost its 'genuine' time-honoured role, i.e. sincerity or authenticity, and culture has itself become a lie as mass culture, propaganda, manipulation and mass media. It is this cosmic scepticism which Ludwig Wittgenstein emphasizes when, by virtue of other conceptual preconditions and linguistic considerations, he describes the unknowableness of the world which is beyond the linguistic medium, the relative difference of the world within the individual languages and the incommensurability of these differences. The disappearance of *Echtheit,* i.e. of authenticity, is not to be measured by the omnipresence of the lie, and the world has not changed into a Crete of logic where everybody is a liar.[4] In the sphere of illusion there comes into being only an ambiguous medium in which each gesture, sentence or even each impulse of expression assumes a dual function. On the one hand it wants to state something and make it public and understandable for others, and on the other hand it wants to change part of that which is to be communicated, hide it, wrap it up in some other context. A constant feeling of defencelessness and at the same time a cynicism in actual possession thrust both thought and then the different forms of communication in this direction, including even the ring of sentences and still later the quite coarse, already recognizable configuration of hypocrisy or involuntary role-playing. Not even children are astonished nowadays at the banal and pious fraud if their mother answering the phone says that 'I am afraid my husband is not in', when he is in fact sitting right beside her. Intention, concealed in politeness and in a 'diplomatic smile' and opposed to its form of presentation, has become a matter-of-course and mutually known untruth and at the same time an indispensable medium

[4] The linguistic analyses of structuralism seek perhaps a compromise which tends to conformity and which is therefore one that is not exactly socio-critical in terms of ideology. They seek a compromise of how it might be possible to employ an exact linguistic analysis in order to eliminate the spontaneous configurations of falsehood which sneak into words and sentences. This language could be a description of plain language, of the artistic form of expression, or the gestures, tattooed drawings and social forms of communication of primitive people (communal contacts, eating habits, arrangement of settlements). The first endeavour was formulated by Sebag [*Marxisme et structuralisme,* Paris 1964] and also by Roland Barthes in his aesthetical analyses relying on the *nouveau roman,* whereas the latter one by Lévi-Strauss in his ethnophilosophy-related investigations [*La pensée sauvage,* Paris 1962; *Mythologie* I—III, Paris 1964—1968], and it does indeed give a clear idea about the way in which semiotics has come into its own as a science.

of exchange. Ibsen was the last in the line of those who revealed what would happen if we dispensed with the 'lies of life' and attempted to live a truthful life without them. The beauty of utopia required only the ruin of a family *[The Wild Duck]*, and the lie of life has become since not only a platitude but also commonplace socially. But in Ibsen's time there were still those who were deceived and took these lies seriously and were therefore shattered both morally and humanly when the untruth of these lies came to light. The lies of life have become refined by now, assuming the character of a medium of exchange which is not only mutually accepted but within which honesty and dishonesty separate from each other. Untruth itself, as a used, somewhat suspected and also invariable medium of communication, has lost its character as a value. It is neither 'evil' nor 'redemptive' but belongs to the sphere of what 'is', and each person discounts its virtuality in his own particular way. The point under consideration is therefore not whether we could exist in our daily routine without these half true and half untrue forms of contact, without the lies of life, but how we can maintain our private or public honesty within this concrete sphere. It is only the assertion of moral autonomy which has become a more dialectical and more complicated process but the bounds of this autonomy have remained the same as before.

The medium of this sphere which is half true and half untrue, but which is in any case in common use, is the inexactitude of linguistic communication, its 'approximation' formed by society, i.e. its ambiguity. On an elementary level the complicity of misunderstanding and lie is not yet intentional. It manifests itself as a characteristic of the linguistic medium, or, more precisely, it is based on this medium, it is entwined with it and forms the imperceptible, half true currencies of social consciousness. Ludwig Wittgenstein called attention to the fact that even our simplest sentences circumscribe the intended subject with exceedingly inexact and indistinct contours, and that all our attempts at exactitude would only dim and confound the meaning more, so it is better to remain with the 'approximation' of standard language. Wittgenstein's example aptly shows how difficult it is to define parlour games, namely where the limit of the individual games lies, to what extent the parties' doings can still be considered a game, to what length we can allow the involvement of fraud, of the spoiling of the game, of alien elements or irregularity (which may still belong among the features of the game), and where the phenomenon of a 'different' game or a non-game begins. [L. Wittgenstein: *Philosophical Investigations,* transl. G. E. M. Anscombe, Oxford 1963, p. 33.] Linguistic periphrasis and definition are nearly impossible because the expressions suggest the essential character of the process only through their ambiguity. But here Wittgenstein, in contradistinction with the *Tractatus Logico-Philosophicus,* one of his earlier studies, knows that this 'approximation', this roughly outlined concept of an object, is part and parcel of the freely tractable normal contacts of everyday life, of its conventions and linguistic storehouse, and that any attempt at 'exactitude' would only spoil these norms which everyday requirements have prescribed. "One might say that the concept 'game' is a concept with blurred edges. But is a blurred concept a concept at all? Is an indistinct photograph a picture of a person at all? Is it even always an advantage to replace an indistinct picture by a sharp one? Isn't the indistinct one often exactly what we need?... For any general definition can be misunderstood too." [L. Wittgenstein: *Philosophical Investigations, Op. cit.,* Section 71.] For everyday thinking and language this approximation and the ambiguity which it covers and leaves open is more important than precise exactitude which is cumbersome for mass usage and which confuses and even upsets certain social relations because this approximation provides an opportunity for statement and concealment, manifestation and reticence, to live in some sort of equilibrium which is conventionalized and indeed ethically neutralized because it is socially authenticated, and to circulate as valid currency and not as a lie, i.e. as a pejoratively tinged moral gesture. Inexactitude and ambiguity are at the same time a self-defence and a possibility for establishing and maintaining contacts. And moreover, this unsureness, i.e. this approximation, is not only the basic form of everyday communication but it is also something that renders help to the development of science. The rigidified and fetishized conceptual apparatus perpetuates a certain structure of ideas, whereas the 'errors' of loose and 'less exact' formulations can lead to further discoveries.[5]

It would seem that his thorough analysis of the structures of everyday language led Wittgenstein to believe that here the point under discussion was not so much an epistemological question (the identity of language and world, language and thought), but rather the necessity of everyday social communication. Approximation is the elasticity of linguistic forms, and it supplies the forms of both expression and communication of everyday life as well as the social and ideological demands with which

[5] "The fluctuation of scientific definitions: what to-day counts as an observed concomitant of a phenomenon will tomorrow be used to define it." [Wittgenstein: *Philosophical Investigations, Op. cit.,* Section 79.]

false consciousness manifests itself. Wittgenstein does not delve into the substance of this wider social horizon; he strictly shuts himself in the immanence of his linguistic analysis and rejects any relationships which transcend language. But his questions still outgrow the bounds of exact linguistics, and he inevitably finds himself in collision with the demands made by society. Perhaps this is why in his discussion with Schlick he explained that the transcendence of these limitations of language is in fact an ethical (i.e., social) question. One always has the need to rush headlong at the limitations of language. Think, for example, of the astonishment we feel when we bump into some (new) existing thing. But this rush at the limitations of language is ethics. [Cf. F. Waismann: *Wittgenstein und der Wiener Kreis,* Frankfurt/M. 1984.] Wittgenstein had a supreme contempt for philosophical endeavours which explored ethical and social relations, because he considered these disciplines the field of ambiguity and indeed of inexpressibility increased to irrationality. Yet his remark shows that he suspected that social requirement which uses approximation to form its own world and hold its own ground against certain ethical deformations was one of the impulsive forces of everyday language. As opposed to the earlier philosophical endeavours of linguistic criticism (which he also practised himself), in his last works he seems to realize that, all things considered, this work of criticism is useless since it cannot eliminate the inexactitude of ambiguity from everyday communication.

To begin with, everyday communication does not need to create and maintain ambiguity or apparency. Without this ambiguity we could not adapt ourselves to the changing functions of our environment and our fellow-men or to the frame of mind of their consciousness. Nor could our manipulation of objects and people be sufficiently flexible. That is why this 'approximation' is the elementary and ontological datum of everyday communication and is to be found even in the simplest construction of our sentences and replies. The girl in love gives expression to her feelings by way of hints so as not to lay herself open, and she relies upon this linguistic possibility just as the shop-assistant at the supermarket does who on being asked for some scarce article does not say 'we are out of it at present', but will list instead 'what is available in the store'. The 'evasive answer', the 'hint' (the ambiguity of which gives at the same time protection against some possible attack), 'veiled suggestion', the 'giving a clue', 'familiarity', 'indirect denial', 'periphrasis', the 'not calling a spade a spade' are all enciphered forms of communication which we use day after day. We no longer notice their ambiguity, and they can function without becoming lies. Add to this the capacity of certain phrases, words and combinations of words to create some peculiar atmosphere, or the nature of the 'depth of text' connecting a different sphere of meaning and perhaps implying a sense which is contrary to the content of the communication. (Wittgenstein already noticed this difference [cf. his *Philosophical Investigations, Op. cit.,* p. 155].) It is through their ambiguity that these linguistic configurations allow of the elasticity of the forms of communication, the 'distanced and tactical nature' of social relations. We could not maintain normal social relations without them. These linguistic ruses which we consider matter-of-course create an illusory sphere which cannot be caught out, but their ontological background is the inexactitude of everyday language, its semiotics which operates in 'approximation'. Nor could the larger units of social life function without them, since in certain delicate situations exactitude would exclude even the possible solutions and thereby make it impossible to avoid the otherwise bridgeable conflicts. (On the political and semantic use of this 'approximation', on its role in the field of journalism and the media see Marcuse's *One-Dimensional Man.* It is the depoliticizing and ideological role of modern linguistics and structuralism with which Marcuse argues here, a role which furthers manipulation.)

Perhaps this is the rational core of the in itself fetishized and absurd thesis which Michel Foucault propounded in his book *Les mots et les choses.* He proceeds from the myth of the isolated man, from the fetishized phenomenon which modern bourgeois society offers, and he calls upon this man condemned to solitude to account for the identification of human existence with 'pure language', i.e. with the form of communication which is devoid of all aproximations. Only a negative result can ensue: pure language and pure human existence which is condemned to solitude cannot coincide. According to him: The only thing that we may know with certainty at present is that existence of man and existence of language have been unable to coexist or to find expression in each other in Western history so far. [Cf. Michel Foucault: *Les mots et les choses,* Paris 1966, p. 350.] The elastic social relations of modern man obviously exclude the demand for a pure language since no such form of communication existed earlier either, and with the complexity of social existence and with manipulation coming to the fore he has become still more entangled in the web of linguistic ambiguities. But it is indeed through the ambiguities of everyday language that this dichotomy of human existence and linguistic existence is cleared up on the level of everyday praxis. The ontological datum of language surmounts this rigid contrast precisely by transcending and maintaining itself, by the structure of 'approximation' which goes beyond itself and yet maintains itself.

For this 'inexactitude', i.e. this 'approximation', operates in a very definite sphere. It does not mean an incongruency (in an absolute sense) between the words and sentences on the one hand and things and relations on the other. What it means is rather an 'indefinite objectivity', one on which Lukács bases one of the key chapters of his *Aesthetics.* Only the semantic range and the outlines of linguistic characteristics can be narrowed down or expanded (sometimes brought to an inverse form), but these semantic changes waver round a core which socially is given with a good deal of certainty. Flexibility on the other hand affords a possibility for things denoted by words and expressions to have a polysemantic, multivalent and ideologically filtered meaning, in which they can join with things and find their way to the consciousness of the other person. I am inclined to believe that when Saussure, the founder of modern linguistics, based his system upon the difference between language and speech, he gave a theoretical expression to this elasticity which is so very indispensable in everyday practice, although he excluded from the realm of linguistic investigation the social context within which in the broader sense linguistic analysis operates. Spoken language is not only richer but also has more meanings than 'the' language, because the concrete situations of the active man (his tasks, environmental influences, active motives, prejudices, ideological attributes) colour and shape the difference and the unity of the objective and human relations denoted by the sentences and the real state of affairs. So we are talking about two planes which shift on top of each other. Both social movement as a whole and ideology contribute to the difference and the unity of these two planes. (It is this socially formed linguistic situation to which Henri Lefèbvre refers when in his argument with structuralism he emphasizes the everyday postulation of human praxis, its mediatedness and 'situatedness' against the fetishized relation between abstract man and language. [Cf. Henri Lefèbvre: *Métaphilosophie,* Paris 1965, p. 265 ff.]) This approximation therefore contains the everyday ontological essence of language. It is a characteristic feature of the expansibility of language, of its ability to maintain social contacts, of people's social practice in which they shape one another. Influence and manipulation are also based on this feature. It is a system of linguistic formulae with reference to the relations between people and things, which in its core is indeed a true relationship but its concealed meaning is the vindication of a social intention. No social consciousness or ideology could exist without this approximation, without the ambiguity of linguistic formulae. The survival of myths, the change of their functions, the semantic change of slogans or the reorganization of conceptual relations are all founded on this approximation which manifests itself in the original linguistic material. (Remember, for example, that the stories in the Bible became the form of consciousness of a particular popular movement by means of the phrases of Luther's or Gáspár Károli's translation, and that neither the consciousness of an age nor even its ideological legacy could be separated from this linguistic form. The elements of myth as regards the blood money or the 'it has been fulfilled' became ideological components in the very linguistic medium in which they had first become the property of everyday communication of the masses.)

That is why an attempt at rewriting, with the purpose of correcting the inexactitude of everyday language by translating its locutions immediately into a metalanguage, seems to be an impossible endeavour. Rewriting means losing precisely the expansibility of the everyday sphere of communication, that is, the intentional richness of 'I hint at more while I say less'. Structural linguistic analysis disregards this broader function of society and language and investigates a fetishized and abstract linguistic form instead of the totality of the social forms of communication. True linguistic criticism must proceed from an analysis which rests on a manifold interpretation of the relations, conceptualizing the linguistic medium as only one of the forms of transformation, i.e. as a necessarily distorting field of the social forms of communication. This might be the way in which this medium could be divested of the unnecessary, ideological or illusion-demanded proportion of the ambiguities which it has itself created.

It is on this ambiguous medium of everyday linguistic communication that all the other half real and half unreal gestures, ethical attitudes and political manoeuvres of social communication are founded. To be more exact, the need of modern society created everyday language from the outset in its own image, framing it into a structure on which this broader demand of society as a whole could be based and which this linguistic medium can indeed hide. It is thus that a seemingly neutral, entertaining or merely average and life-sustaining ideology entwines with the ambiguity of linguistic formulae and other semiotic systems. Everyday language is dissolved in everyday ideology, just as everyday ideology is also made up of many linguistic factors such as light music, the semiotics of mass entertainment and the stereotypes of mass communication. Modern life has brought about a change here, too. The quasi-reality of everyday ideology exerts an influence today from beat music through the transposals of various fashions and behaviour patterns to the habits of speech and prescribed forms of thought, and compels a man by the speed with which it spreads, which is more rapid than was the case with similar kinds of everyday ideological formations of earlier periods, to accept these contents. Hit-tunes of thirty years ago were in vogue for many years after their time; the legend of Sonny Boy, helped by time, custom and association with it, figured as a substitute for everyday reality. Today the situation has been reversed: it is precisely through their fast rate of change that the everyday vehicles of ideology compel a man to follow the current trend and to forget to think independently in this hurry. The quickness of change has become an instrument of suggestiveness. On top of it all, this feverish material exchange occurs in a linguistic material which cannot even be translated without mediation. It prepares an inner disposition rather than a perceptible brainwash. It simply makes us susceptible to certain ideological trends and renders us defenceless against manipulation which assails our integrity. And that is all it wants. And while rushing after the changes of this medium we do not even realize that, although surrounded by the realities of everyday life, we have been moving in an unreal medium long since. What has changed is only the way we look at things and with it the semblance of everydayness. Under the circumstances it is quite impossible to decide who is a liar and when he is one. That is partly because this lie has become so very impersonal while its usefulness, its role filling out our life, its entertaining quality or its known and suspected ambiguity so plausible that it has become a part of our life. The rise of the Beatles and then of the more and more recent groups succeeding them, the alternate succession of the highly coloured psychedelic conditions which they created, have collectivized this unreality. As a result, unreality has not only become depersonalized but has also been raised to the level of plausibility. There arises a scintillating realm of ambiguity, one which is difficult to break through in view of the 'closed order' of everyday spontaneity. And the endeavour of everyday ideologies is to maintain this inclusion. The fast pace frustrates one's attempts at becoming independent. There is always a new style that we must overtake or, in the simplest case, we are supposed to know about, and with this we are at once tied to the sphere of immediacy: the realm of ambiguity.

It is thus that the complicity of the lie and misunderstanding or their imperceptible team-play is brought about. The lie dominates in an adult and cynical form, whereas misunderstanding does so in childish spontaneity. The lie of the 19th century was still an enormous, career-creating bluff which founded the rank and the wealth of the Rastignacs and which needed a 'superintelligent' performance of a manoeuvre of false representation such as we encounter today only in the better crime stories. It was not enough for one to know by heart the relentless, steadily and minutely elaborated scenario of hypocrisy and role-playing, but one also had to possess the asceticism of lie. A cold flat, a furnished room and squalor were in store for Rastignac (or Raskolnikov) on his return from the glamorous salons, and the next day he had to carry on with his double life with the same obsession fuelled by resignation. In most segments of present-day society the scenario of role-playing is played as a matter of pure form, and the individual steps offer themselves of their own accord. What is difficult only is to take some other steps, because the rejection of everyday ideologies would mean giving up hypocrisy and would require asceticism, moral autonomy and independent reasoning. And this is the point where I want to emphasize the attribute 'everyday', for indeed what this category means is not identical with the 'official' ideology of a particular society. In our case this everyday 'nonconformity' does not mean that someone denies the tenets of Marxism at home... The medium which puts one in bonds and is hard to repudiate is a much simpler one: it is spontaneity, the reign of immediacy, an ideology which has been latent in the structure of Philistinism for centuries on end, i.e., the ideological sphere of everyday life which has never been illuminated 'officially' within the four walls. It is difficult here to assume a nonconformist attitude, for one must in the Goethean sense really fight for freedom every day if one wishes to live a life worthy of man. Spontaneity corrupts like morphine. It offers a narcosis which disconnects the circuit of the daily rat race, the struggle of public life, drudgery of labour, and at the same time turns on an ideology which has an imperceptible way of stupefying but which renders itself indispensable through its soothing effect. Whereas in the 19th century one needed asceticism for the relentless accomplishment of a lie, today misunderstanding plays into one's hands. The submitted and embellished reports and their 'well-intentioned' or equivocal acceptance have dealings with the basic questions of philosophy and ideology only at a distance; they in fact follow the structure of everyday ideology, the unconscious team-play of half-truth and half-misunderstanding. The point is that in this team-play there is always some grain of reality which factually substantiates that which is added to it, which sticks to it as untruth. And it is this grain of reality which furthers (morally too) the mechanism of misunderstanding, the gesture of self-deception which accepts illusions and reassures, in fact sustains, itself. (As we have already quoted, Felix Krull knew that an effective lie was only one which included at least a grain of truth in the story.) This is today one of the basic axioms of the team-play of half-lie and half-misunderstanding, a principle on which the creation of everyday illusory sphere depends. Outright lying is easily detected even today. The only lie for which there is no real antidote is the one which is charged with an element of reality or which has assumed a function of reality, particularly in the case when misunderstanding consciousness expects a remedy for the effects of a lie in the form of morphine. The man who lives and works in immediacy will himself seek any possibilities of misunderstanding that are on offer, for they provide him with a moral loop-hole to shirk his responsibilities and excuse him from the recognition and the shouldering of reality which is pregnant with unpleasant consequences. (A mother will not only turn a blind eye to her daughter's misconduct, but she will in fact believe her flimsy lies up to the point when their team-play weaves a fabric which cannot be undone any more from either side. A more involved procedure will undoubtedly ensue when the boss does not give credence to his employee's excuses for his conduct, but here, too, a misunderstanding intercalated will be of avail.) A different form of misunderstanding is when someone covers himself with the moral anticipation of success. This is true self-deception since for the duration of one's belief in success one will handle as reality that which is in fact only a supposition, an intended illusion. (Gyula Illyés gives an excellent description of this character in his *Kegyenc* [The minion]: it is on the evidence of the future that Maximus bases his 'different view' of the present, the misunderstanding and self-deception through which he can accept the necessary existence of his despot. On top of it all, this is a puritanical and ascetic self-deception, for the reality of the sacrifice covers, as it were, morally that which does not even exist in reality.) Hence misunderstanding and falsehood in the short term become accomplices or the kind of partners which are unconsciously complementary to each other in a false reality, since within this distance this symbiosis turns out a success. Here the result is also true besides the moral sacrifice. And in retrospect the factual contents which misunderstanding and falsehood have mutually established become real.

The fact that misunderstanding and lies have become everyday occurrences and that the moral and ethical aspects of their value have vanished is to a certain degree due to the minor and inevitable wrongdoings of everyday praxis. By inevitable we mean that by virtue of the ontological nature of everyday life we are compelled to commit minor transgressions so as to avoid the ones which are considered grave from a moral point of view. The imperceptible, unconscious or self-deluding complicity of misunderstanding and lying in dropping the matter of these 'minor sins' is necessary for maintaining the alienated mechanism of everyday praxis. They are almost automatically deemed 'not to have happened' because their inevitability becomes a routine conflict after a while and yet ethics does not provide it with a 'special' release clause. Consequently in the everyday sense there is also a *'bona fide'* version of misunderstanding. It is no mere chance that, as a spontaneous illusion-creating gesture, this 'good faith' is itself associated with the history of capitalism. The essentially historical conflicts which are inherent in the contradictory nature of *'bona fides'*, the rascally acts done in good faith, are first raised by Schiller in his last great drama, the *Demetrius*-fragment [1804]. In feudalism good faith was a natural and observable form of duties of an order or estate. The sphere of existence composed of the dichotomy of the objective practice and consequence separating from man on the one hand, and of intention, purposefulness and goodwill on the other hand arises in capitalism where the result of an action ultimately loses touch with one's intentions and turns against one in an alienated form. It is this contradiction which Schiller formulates in a historical tragedy. But with the consolidation of the fetish-like character of capitalist everydayness the historical order of magnitude becomes trite. The good faith of the little man turns into his being deceived, into the comic blindness of fetishized consciousness. All the more reason why it is difficult to break the alliance of misunderstanding and lie. With the consolidation of the fetish as a character it also becomes more difficult to recognize this alliance for what it is, and 'good faith' becomes itself an ambiguous excuse which is objectively negative but subjectively acceptable. A break of this alliance takes place only in revolutionary moments or in the case of moralists who are filled with revolutionary ardour. When it comes to that it will meet with success in the former case and with the risk of failure in the latter. But these 'minor sins' round which misunderstanding and lying build up the everyday sphere of illusions (a pretence that 'nothing has happened') form a category which can considerably be expanded and which is in fact devoid of either objective or subjective limits. That is why the concept 'wrong' — which is necessary for everyday life — may include the kind of humanly degrading acts which lead to the final alienation, to moral disintegration. As we mentioned it before, the reason why the Catholic Church introduced the absolutization of contrition was because they realized that in everyday logic dropping the matter of 'minor sins' constituted a matter-of-course condition of survival, or to be more precise, of continued vegetation, but the moral echo of this not only stays with a man but can also be exploited to the end of maintaining spiritual and intellectual control. Confession, i.e. the transcendental dependence on sin, makes it impossible for a man to be released, and the new 'everyday' sins will only consolidate this state of dependence, this spiritual and ethical control. By doing away with general repentance Stalin's era broke the bonds of sin and thereby rendered redundant the periodicity of the succession of confession, absolution and new sins. Stalinism took as its starting-point what followed automatically in some form or other from the logic of everyday life: everybody is guilty, whether because one believes that one had to do something which was indeed necessary but which morally counts as a crime or because a tactical step was frustrated by one's moral resistance and one thereby offended against the discipline of the movement. So it is precisely for the sake of maintaining this control that it is not worth the trouble of insisting either on an acquittal or on the pronouncement of a guilty verdict since invisible guilt which can only be suspected sets things in a stricter order. Thus there emerged a 'wily fear', a psychosis, in which the innocent could believe in their guilt because even those steps of theirs were taken into account in their consciousness — consciousness in the everyday sense — which incidentally were declared 'not to have happened', or because in certain historical periods such as war, intervention or at the time of building socialism in a particular country their true moral autonomy began to give way under the weight of the relatively just requirements of discipline. In this era the complicity of misunderstanding and lie broke only a very narrow trail between morality and legality, although morally this narrow trail turned out to be rather wide: the latter produced not only careerism and immorality, but also 'gross manipulation', the historically tragic and discreditable illusory sphere of a radically new era.

The need for misunderstanding helps to develop the ambiguous formation of language in which anything can be stated in such a way that the gist of the statement cannot be proved but that he to whom it is directed is wise to it. One can 'read between the lines' in the newspapers; the protocol enumerations rated as the communiques of the political state of affairs for a long time, and 'functionary language' also served for fulfilling the needs of this ambiguity. The expressions 'it is no mere chance', 'we must go over the matter again', 'the issue has to be interpreted in political terms' conveyed directives whose content was entirely different from what the un-Hungarian etymology of the words meant, namely overt or covert summons for administrative measures. And this linguistic fabric formed the political sphere in which unreality and reality were nearly inseparably entwined, already at the source and in some concrete case at the end of linguistic expressibility.

But it is not only the political sphere which can create this ambiguous linguistic medium. As a matter of fact, it only makes independent the linguistic structure the origin of which is the order of communication of everyday life, i.e. the common ambiguity which has been developed by the elements of our sentences used in awkward situations and by the 'narrative structure' that we use in covering our misunderstandings, our self-deceptions or our innocent lies. When the everyday sequence of sentences is saturated with the conditional modes of 'perhaps', 'if' and 'although', when sincerity can be a different form of lie, this will give rise to consequences which come near almost to the roots of grammar. An investigation of the development of everyday language might reveal that the linguistic reign of this ambiguity which has become matter-of-course extends all the way from intonation to word order. Here are a few examples to illustrate the double-bottomed use of this concept. Consider the increase of superlatives which reflect the banality, the triteness and the familiarity of bare facts, which are disheartening even in terms of their own value, and that against this a linguistic form had to be found which was seemingly more awful but in its shock-effect 'more facilitating'. Or there is the syntactic construction which is tinged by the uncertain certainty of 'somewhere': 'This writer is right somewhere.' The adverb of place indicates that it would be disagreeable to state where he is right and about what, so it seems more prudent to remain in the fog of ambiguity. 'I will drop in on you one of these days', says someone when coming across one of his acquaintances in the street; in fact this means 'leave me alone', but 'one of these days' is an ambiguous locution since it may indicate an indefinite time but, provided that we make a polite interpretation of it, may mean even the next day. It is likewise a matter of common knowledge that sentences beginning with 'one' usually indicate first person singular but its ambiguity affords the opportunity partly to conceal that it is me who I am talking about and partly to generalize the individual contingent case.

This linguistic ambiguity is not yet an illusion. But as the objectified form of the expression it helps to build the illusory and to break down the real which is hidden anyway. Ambiguity has therefore a bias: one meaning sustains that which creates illusion illegitimately, whereas the other maintains that which has been refined into legitimately employable so as to create conventions. Neither are these lies in the 19th-century sense of the word. They are only 'locutions' out of which the everyday forms of our communication are constructed. And since they become matter-of-course and indeed indispensable requirements, it is increasingly difficult to break through the fetishization of their linguistic forms. For behind linguistic formulae which have been objectified and made into clichés, i.e. used by all and sundry, there is a mass of human relations where even language is highly responsive to the collisions of distortion, fetishization and reality. It is this conflict which the ambiguities of language give expression to even in the case when they are biassed.

Untruth enclosed in the clichés of sentences of spoken language has become accepted as natural because a much cleverer illusion-creating medium has also come into being, namely linguistic content conveyed by objects and things. Modern manipulation exerts its influence not so much through the communication of newspaper, radio and television as by means of an ideology which is built into objects. Manipulation produces a conceptual position in which one will read and listen to one's neighbours and society in a different manner. It is of course undeniable that, as ideological vehicles, all amenities of comfort are ambiguous in the same way as the prescribed forms of speech are. The ideological content of their answer depends on the manner in which they are approached. The ideological content of the car, the family home, the comforts of life as well as their grammar change according to the way in which one approaches them. There arises a diabolic interaction in which one does indeed shape one's own environment and elicits an answer from the surrounding objects by one's approaches, but the 'ability of one's approach' is itself shaped through the language which is programmed partly by objects and partly by everyday ideology.

4.4. ARTISTIC ILLUSIONS AND THE ART OF ILLUSIONS

'Work of art, time and illusion all mean the same, and together they fall victim to criticism which can no longer tolerate illusion and play, fiction and the arbitrariness of form, that is to say the autonomy which has subjected passions and human sorrow alike to its censoring and which has assigned them parts and projected them into images. What is permissible is only the non-fictive, undistorted, unplayed and not even cleansed expression of passion ... It is the illusory quality of bourgeois art which falls victim to criticism.' Contrary to all appearances, it is not a Marxist writer but the Devil himself, Samiel in Thomas Mann's *Doktor Faustus,* who utters these words. His summary judgment according to which the entire bourgeois art is one of illusory quality is just as much of an exaggeration and infernal dogmatism as it is erroneous, for indeed every art creates its own particular sphere of illusions. And it depends on the content of the work of art and on its realistic power of evocation whether this illusion is 'well-disposed', in other words of the nature which reveals our true world and awakens the recipient to himself and also to his historical present or, on the contrary, it is 'ill-disposed', insofar as artistic illusion becomes an accomplice of untruth which has been depicted but whose illusory nature has not even been queried. For all that, there are many elements in this diabolical logic which give food for meditation. First of all, it delineates the programme of modern art as an immediate, unarticulated expression of human sorrow, of suffering which cannot be given any form, in opposition to the earlier (classical) endeavours where the artistic formation and articulation of expression also gave rise to the illusory, to the untrue and to the factitious. In Samiel's words the articulation and the illusoriness of art are entwined with each other; both have to be rejected as mutually destructive principles. In other words, he condemns both what furthered the lucidity of reality in the development of art, i.e., the artistic restoration of the human, ethical, relational and formal articulation of the world, and illusoriness itself of which Samiel's interpretation is probably that any mimesis can reflect only a delusive appearance, a false veneer. When he rejects the illusoriness of bourgeois art he does not aim at doing away with false reflection or illusoriness which accompanies and influences the development of the art of the period, an illusoriness which had a disturbing effect on the correct picturing of reality. What he has in view is that one must break indeed also with picturing itself, with those forms of the illusory world of the work of art which give expression to reality. Samiel states the aversion of the modern avant-garde artist, an aversion which wants to break with every pictorial thing, with a practice which uses the surface elements of reality as a means of expression, i.e., with the anthropomorphizing artistic method. He wants to reject both the conservation of false illusions and the anthropocentric method of picturing which forms the essence of art, the decadent deceit and the 'deception' of art in revealing reality.

With this, however, we have at once reached the key question of how the avant-garde conception of art is related to illusion. We propose to formulate our question as follows: in what respect can we find some sort of an affinity and relationship between the decadent trends and the world of illusion? We are namely of the opinion that the attraction of decadence means after all that it is the surface illusions which determine the ultimate scope of representation.

Round the turn of the century the answer was much easier to find. Illusion was also more highly valued then, in the 'heroic' period of the development of decadence. Nietzsche's philosophy of art expressed this elated attraction, the allaying of qualms of conscience with regard to illusion as follows: 'The lie is justified in art, and the will to deceive *(Wille zur Täuschung)* may rest satisfied with the knowledge that it has a clean conscience on its side.' Thomas Mann, in one of his studies on Nietzsche, even points out that the *Weltanschauung* of this idea has so much in common with Oscar Wilde whose literary works now seem second-rate but who was nevertheless a very influential thinker round the turn of the century. He quotes Wilde as follows: 'For however much we keep trying, we cannot gain access to that which is behind appearances, to reality. And the horrible reason for this is probably that there is no other reality than the phenomenality of things.' From a more conservative point of view and with less frivolity the theory of 'empathy' also registers this conceptual attitude. Lipps formulates this relation as follows: With this (i.e., with 'empathy') there seems at once to be given a typically immediate intellectual understanding; with my eyes I can see immediately what the forms mean. But there is also more in seeing, namely empathetic perception. [Cf. Th. Lipps: *Aesthetik,* Hamburg 1903, p. 109.] Lipps of course does not draw the conclusion of this thesis which thus substantiates the illusoriness of reality, because art and artistic pleasure are compelled to have recourse solely to the illusions of immediate observation. But this conceptual attitude must have been so much in the atmosphere then that it was adopted without a hitch by other, more sedate, theories concerned with aesthetics.

At the turn of the century, and even more so in the crisis of World War I, a different and contrary artistic movement gets under way, one which is controlled by a lack of trust towards the immediately perceptible. Already Cézanne notices the discrepancy between the illusory and the real within the perceptible itself, but he wants to resolve the argument by making the visible world more stable and more substantial, by seeking for its social and human content. In a discussion with Gasquet he says: Everything we see is unreal, scattering, and escaping us. Nature of course is always the same but there is hardly anything left of its phenomenal picture. Our art must reproduce the grandeur of permanence. What is there behind natural phenomena? Perhaps nothing, perhaps everything. [Paul Cézanne: *Über die Kunst,* Hamburg 1957, p. 9.] Later this attitude becomes void of excitement and there remains only a lack of trust towards reality. Reality has vanished and we live among only illusions. It is in this connection that Bahr and later Ernst Bloch, the theoreticians of expressionism, attack the illusoriness of the bourgeois art of their period, the yielding to illusions. Even Brecht's fight against the theatre of illusionism can be attributed to this movement; the configurations of immediate reality can be received only with intellectual criticism, and art must not be a naive recreator of the surface meaning. It is easy to see the two-facedness of this phenomenon. It is on the one hand controlled by the pathos of social criticism. It wants to reject the illusoriness of the surface of the average bourgeois life because it senses that the immediately given world is false and so it must go back to the deeper layers in order to expose this world. It is this line along which the lack of trust towards reality also protests against the reign of kitsch and commercial art. For when kitsch acquires a dominating role not only art is to blame. Life itself creates kitsch.[6] We must therefore turn away from the false world of immediate reality. But on the other hand this gesture also implies the fatal misconception that art at the same time forgoes its basic means of expression. The turning away from the immediacy of reality also means that art denies this unmediated sphere in the form in which it is recreated by expression and reflection. With this the originally just aversion assails art's basic medium of reflection. The only way in which essential relationships can be humanely represented is if they are made to glimmer through the surface appearances and illusions. But the lack of trust towards that which is perceptible and the fact that immediacy had fallen into discredit gave rise at the same time to an estrangement from the entirety of reality and to the complete denial of the immediate surface. The illusory cannot be broken down and then restored in its exposed form because the building material spreads the contamination. There remains the 'deeper' layer of reality, the cold reserve as regards surface elements. It is also this relationship, the absolutized criticism of illusoriness, which is behind the endeavour towards the 'pure' arts. All surface elements which are perceptible — even exposed illusions — produce illusoriness. This is one

6 Kitsch could not have been created and could not be reborn day after day if it had not been for kitsch-people who like kitsch, and who as art-consumers are also inclined to buy it. Art is always the image of the man of the age and if kitsch is a lie then this reproach is also due to the man of the age who has a need for such lies and beautifying mirrors. [Cf. H. Bloch: *Dichten und Erkennen,* Zürich 1955, p. 295.]

of the sources not only of the abstract tendencies in painting, music and lyric poetry, but also of the slogans of objectlessness, etherealization etc.

Of course, great artists realized the extent to which this slogan which was progressive in its purpose was undermining art and so they desperately resisted this trend. They wanted to grasp the phenomenal sphere, the elements of the immediately perceptible world, in such a way that their illusoriness would be revealed in this immediacy, in the world of exposed phenomena, and that at the same time their true relationships would also glimmer through. We have already mentioned Cézanne's endeavours at trying to restore the picture of the world and to orientate himself in reality. Van Gogh, another great master of the age, pursues a similar course. He paints a café, i.e. a 'place of amusement' in the sense of immediate illusoriness. But the way he does it is that the message of his painting is in dramatic contrast with illusion, with the falseness of gaiety. In a letter to Theo he writes as follows: "In my picture of the *Night Café* I have tried to express the idea that the café is a place where one can ruin himself, go mad or commit a crime. So I have tried to express, as it were, the powers of darkness in a low public house, by soft Louis XV green and malachite, contrasting with yellow-green and harsh blue-greens, and all this in an atmosphere like a devil's furnace, of pale sulphur." [*The Complete Letters of Vincent van Gogh,* London 1958, Vol. III, p. 31.] The phenomenal elements of the perceptible world thus serve to expose the real world in an intensified arrangement which is artistically reinterpreted according to the essential human relations. The phenomenal is rescuable in this sense. The effort to maintain a this-worldly view of artistic interpretation (Mexican painting, recent graphic art in Italy) is only a counter-trend, the main tendency being the critical disintegration of the perceptible world and its desperate or hedonistic denial: the absolute identification of illusion with the perceptible and the rejection of the perceptible. With this, however, art's medium of expression, the sphere of phenomena conveying and representing human relations, will also become a dubious instrument and, moreover, today even a downright taboo.

Avant-gardism is a two-faced ambition with respect to illusions, too, since it realizes that the illusions of bourgeois society breed in the immediate phenomena of life. It was due to this powerful attraction which this critical element of avant-gardism had for progressive art of today that many artists who are now socialists have reached socialist realism (Nezval, Aragon, Brecht etc.). But at the same time the complete denial of the phenomenal elements of immediacy and the break with anthropomorphic illusions gave rise to a whole series of inconsistencies which have consolidated and maintained the decadent tendencies, sometimes without the will of the particular artists. The turning away from immediate reality and the representation of the purely 'spiritual' essence can come to fruition only if they still retain some randomly selected and independent element of the immediately factual, and in its allegory, i.e., in the mirror of the perceptible facts, they still throw light on the imaginary or true essence. What do we have specifically in mind? One of the most important social-critical facts of the absurd drama of today is that people do not understand one another and that they are indeed unable to do so. Each person is confined to the prison of his own alienated life, from where he can make only indecipherable signals. But the avant-garde theatre proceeds to tear out of the web of contexts this superficial fact which is true as an element and which is symptomatic of the atomization of the bourgeois society of this age, and represents it in a cosmic enlargement as the ontological damnation of man's destiny. And this absolutization obviously also includes the pathos of the critical protest against manipulated gaiety according to which everybody gets on well with everybody else, a gaiety in which the society of smiling people happily tell one another the boresome nothings which the hit-tunes, sob-stories, TV-dramas etc. have put into their mouths. Ionesco's *Chairs* is not the solution of the social problem; it rather shifts the solution onto the level of mythical relationships. The phenomenon becomes the explanation of itself, namely the explanation for the ingrainedness of incomprehension, and there seem to be no possibilities for a solution; it can denote only itself — and this is the essence of the world. But this thesis which is in argument with immediate illusions leads itself into illusions, because the manner in which this 'phenomenon' holds sway over people is by no means so mythical. So the denial of surface illusion results in the affirmation of an illusion which seems to be more profound but is in fact just as shallow.

The artistic method of this contradictory relation and at the same time the technique of developing an illusory sphere of mythical meaning is the absolutization of partial phenomena. Fear, for example, is a real fact of life for the European man who has lived in the midst of world wars, reactionary dictatorships, fascism and the cold war. But the only way in which musical compositions, dramas and artistic endeavours of painters will realize their belief according to which they can find the 'essence of the world' solely in this element is if they enlarge this real fact to a cosmic magnitude, that is, make it illusory. Enlarged to a mythical magnitude this fact will hide genuine relationships such as, for example, the play of forces which are aligned for and against war, the hidden dialectic of the controllability of horror and of its getting out of hand, the involute order of the up and down, visible and invisible relationships, so the mechanism which made the connection between the immediate surface and the 'silent depth' more complicated during the past fifty years (at the time of the most recent stage of development of decadence) than ever before. When this cosmic fear becomes the principal message and the creative force in Beckett's plays then its articulated order, as it manifests itself in real life, will not be perceptible. Each person will read into the mood of the inarticulate dread of the dramatic picture whatever he wants to. (It is no mere chance that the mythical explanation of *Godot* already has its own literature; it can mean God, Bolshevism, fascism, the hereafter etc.) What Samiel suggested, namely that the artist should express human sorrow in its own inarticulateness, has come to realization here. But inarticulateness also means that the concrete features of man and the world have vanished and that instead of a face only a blurring mask can be seen: there is only a single phenomenon, which has a like interpretation for everything else, for every articulated relationship and hierarchy, for the *explanans* as well as for the *explanandum*.

The technique of the absolutization of partial phenomena is not really a new thing, of course. It is novel only in its propagation, only in its becoming a universal method. The novelty of Cubism was to express reality in cubes, in geometrical elements, i.e., to reduce natural forms to their geometrical equivalents. In this endeavour Cubism cited the authority of Cézanne who was the first among the painters to formulate this abstract method. But with Cézanne the 'circles, squares and planes' were still of secondary importance. They helped to establish the social and human totality of reality. Cézanne says in fact that he wants to paint the landscape the way in which the farmers see it, and he raises the problem of geometrical frameworks as one possible way to express this visual perception, hence as a technical element of secondary importance. But there are no triangles or squares in his paintings, though we can discover them in the contours of the marble slabs of Mont St.-Victoire if we look for them.[7]

[7] Cézanne writes about the artistic ideal he strives for as follows: The peasants know what they must sow here or there, what the weather will be like tomorrow, whether there will or will not be a cloud-bank over the summit of Mont Sainte-Victoire etc. Just as the dog knows about its food so they also sense these things and orientate themselves solely in accordance with their exigencies. But I believe that most people do not know but only subconsciously sense that the tree is green and that this green is a tree, and that this earth is red here and that this reddish brown hummock is a hill. Without losing any of my present knowledge I ought to attain this intuitive mode of perception, and the colours which are scattered over the ground here ought to have the same meaning for me as they manifest themselves for them in harvest. [Cézanne: *Über die Kunst, Op. cit.*, p. 21.]

And indeed, he predicts to Gasquet that his remark on squares and circles which with him are a methodological aid to a richer way of seeing will later become a fashion, similarly to the method of impressionism. [*Ibid.*, p. 22.] Cézanne already saw that the absolutization of partial phenomena had begun. And his prediction has indeed come true: Cubism simply disregarded the mode of perception which wanted to see the landscape as seen by the peasants who lived there, and emphasized instead one element of that landscape, namely squares, geometrical elements. And with this, Cubism set a screen before reality, bringing about the loss of the abundance of depth and varied aspects, i.e. the total richness of the scene which, with one last spasmodic effort, still vibrated in Cézanne's paintings.

So the turning away from reality is compelled time and again to return to surface phenomena for a particular element, in order to generalize this surface element and construct from it a mythical and theoretical world which bears no resemblance to everyday illusions but which again has some leanings towards illusions of a different structure. This theoretical type of the illusory world is the unstable field of modern art and its tempting possibility of expression. A great many things can be said with these allegorical forms, and the reason why the painting created in this way can indeed have a tremendous impact is that it can express the common essence of numerous phenomena. Yet precisely because it does not differentiate among the essentially different things but reduces them to a common denominator, i.e., subordinates them to a common surface element, this highly effective general picture remains illusory. And the essence of this method is that surface denial can be realized only by maintaining the surface elements. What is rejected as a whole is brought back in details as a means of expression. Tristan Tzara's slogan that 'objects must be placed beside one another in accordance with the laws of chance' expresses this principle still with a playful unconventionality, whereas Lautréamont sees the denial of surface facts already in their arbitrary and mythical re-creation: 'Beauty is when a sewing machine and an umbrella accidentally meet on the operating table.' For if the surface structure of reality is the feigned and contrived lie then art may be total uncontrivedness, i.e., irrational and aleatory. But everyday surface events are assembled into some 'banal beauty' in the pages of magazines or in average art just as arbitrarily as they luxuriate in the wild contingencies of dada.

At a later stage this development starts out from more intricate phenomena of life. A particular fact of life is selected from reality and absolutized. The basic mode of perception of a particular work of art and at the same time its perceptible and enjoyable content and action are given by fear, loneliness, dreariness, incomprehension etc. These facts are the elements of reality of the bourgeois weekdays. They are not invented elements such as the contingency of dada. Since, however, they stand by themselves, having been taken out of the fabric of their contexts, they become a mystical magnitude and consequently illusory. The method is modified here by e.g. loneliness becoming at the same time a phenomenon of the life of individual heroes and the keynote and content of the work of art. It is simultaneously the individual man's destiny and a key to the world. The individual fact of life will be immediately identical with the essence of reality. This is the destiny which in Sartre's *No Exit* confines the three people together, and only this may account for the nature of the allegory, his innuendoes which he lets fly at the world. Here, too, the mechanism of the ideation of illusion is at work; we can see a gradation of analogical reasoning which uses a particular phenomenon simultaneously as an individual concept and as a universal formula of interpretation. It is by making the essential relations glimmer through, i.e., by means of displaying the total and manifold richness of the inner hierarchy of the work of art, that the realist artist can show the illusory and the real relationships of the immediacy of the surface, and can elucidate these phenomena, expose the illusions and make a proper interpretation of them. Even in the well-meaning protests the influence of the attractive force of decadence takes effect through immediate facts becoming at the same time, and immediately, explanations of the essence of the world. The artist comes under the demonic influence of this attraction when he loses his autonomous ability to analyze his relation to these surface facts and handles them as immediately perceptible essential facts.

That is why I think that it is a misunderstanding to suppose that modern art strives for immediate expression of the essential in the same way as the scientific world picture of the atomic age has made the essential a fact of common knowledge for modern man. This is an illusion carried by theories which see in abstract painting the expression of the geometrical perception of essence, in pure poetry the grasping of the law-determined 'spiritual' content which cannot be put into words etc. According to Apollinaire, the mission of 'pure painting' is to reveal a drossless reality which has been cleansed of illusion and to return to the idea, to order, to the law. We can still hear in these words the protest against illusions and the striving for the essential, yet it is this slogan which leads to the absolutized use of the elements of the everyday surface. What he believes to be essence, i.e., what is beyond the deceptive sphere of immediacy 'covered with dross', is in fact only a minute particle of immediacy. The triangles and circles, the pure sounds, solitude, incomprehension etc. are all immediate elements of the perceptible world, only in the disguise of a formalized abstraction made into a cliché. As Jean Bazin, the French essayist with an excellent intuitive perception, writes in this context on Surrealism: If therefore a surrealist painter makes careful copies of the pictures of a particular dream, then this method amounts to no more than merely copying any other object. Dream is nothing but the senseless reality of everyday life. It is not a different reality. [Cf. W. Hess: *Dokumente zum Verständnis der modernen Malerei,* Hamburg 1964.] The allusion to everydayness aptly indicates that even in such a form of transposal surface immediacy and illusoriness do arise and that it is by rearranging their elements and divesting them of their essential connections that the artist forms his paintings. But the senseless elements of the everyday world remain illusory even when we remove them from their false forms of presentation and mount them on the canvas of a new — contingent — arrangement.

It is thus that we see the theoretical categories of illusion again. In the first place, the artist presents reality through a single point of view. Avant-gardism is drawn into the dangerous enchanted circle when it wants to maintain this 'single viewpoint' mode of representation ever more forcibly: the world can be seen only as a 'case' of alienation, solitude, communication disorders etc. Whereas the realist artist seeks totality even today, making the destinies he represents accessible for an approach from all sides and trying to display every aspect of them, the effect of decadence is that artists are impelled to present the totality of a single viewpoint. Dürrenmatt, for example, is a realist when he starts out from the totality of relationships in his play, *The Visit of the Old Lady.* The events that took place in the little town of Güllen did not occur on account of the power of money, nor even because of 'the rule of conformism' or revenge or the victory of the subconscious. They occurred in the web of the effects of all these and other factors which partly thwarted and partly advanced each other. This drama is not the allegory of a single relationship but the parable of many. When we are left only with a single way of approach to phenomena, when it is the psychological mechanism of conformism that Ionesco creates a dramatic and gradually increasing tension in his *Rhinoceros,* then we can apply a variety of interpretations to the picture this gives us. It depends on our individual disposition whether we interpret it as the development of fascism or as the criticism of communism. And, since the one-sided approach encompasses the world in a single explanatory cliché, we recognize only one of its projections as its essence. This, too, is a matter of the modes of perception which have developed on the basis of certain stereotypes and by which we 'must' see reality without being able to gain access to its essence. This picture is then very exciting, interesting and fascinating with most of the talented artists of the avant-garde. But its oddity and interestingness is indeed the very pitfall in which both the artist and the public are in thrall to illusion. Oddity is a category which furthers the propagation of the clichés of illusion and its binding nature. The invention of newer and newer and ever more cunning interestingnesses has become a requirement which impels artists to introduce a particular angle or story line because it seems interesting. Interestingness has acquired independence as a means of manipulating the public, and against the interestingness of yesterday one has to create the oddity of today. This is the requirement that does indeed control the development of clichés. When linguistic incomprehension came into vogue it was this incomprehension that was interesting and it was this formula with which one had to work. Kafka's influence functioned similarly in the hand of his epigones. A particular idea (such as, for example, an endless tunnel, a train which never stops, a house in which the tenants move ever higher and somebody dies on each floor etc.) becomes immediately a principle of form and can be raised into the work of art as a prefabricated component, and here, in the work of art, it will from the outset determine the attitude of the characters, their destiny, their psychology and their *Weltanschauung*. Just as in the average life it is these ready-made formulae for living that determine the microcosm of people (fashions, types of behaviour, gestures, acceptable standards of conduct in politics and public life etc.) so these formal clichés determine the structure and inner order of works of art. Curious though it is, but the fact is that only kitsch still works with such off-the-peg technical recipes. The sickness of avant-gardism commences when it is unable to achieve originality in a deeper sense and is forced to maintain the illusion of interestingness solely by the originalities of the recipes.

That fraction of modern art which has got into the danger zone of decadence works with more abstract, more theoretical illusions than did the similar endeavours round the turn of the century or even in the twenties. It confronts with stronger contrast everyday illusions characteristic of 'magazines' and creates its illusions on a more abstract level, deriving them from critical elements. Yet these formal problems of representation simply hang in the air if we do not embed them into an invigorating *Weltanschauung*. The absolutization of partial phenomena is not an accident. Insofar as, for example, solitude becomes the sole formal and contentual manner of representation, this endeavour is also maintained by a conception of the world. When an artist can see through the immediately effective play of inhuman forces and surmise the seats of resistance, he will also notice that these partial phenomena are particles which have been tragically arrested in a greater movement but that they are on their way toward a solution. Decadence can exert its influence where the artist regards alienation as a normal condition, where he accepts, either bitterly or with a natural cynicism, the existent world order as the law of 'the' world. It is the world concept of resignation which prompts the feeling of indifference and impotence. Decadence is therefore the poetry of paralysis, and it manifests itself only where the world concept cannot step out of the mood of resignation.

This struggle between resignation and resistance, paralysis and activity within every art, artist or even within single works, has not been resolved. Artists recognized as decadent may come forward one day or the other with a work in which they radically break with their hitherto developed conception of the world, and on the other hand progressive artists come under the influence of decadence when they get entangled in the inconsistencies of a particular work of art. (A case in point for the former is Adamov and for the latter Dürrenmatt.) Accordingly, it is hard to put one's finger on any unified set of phenomena that we might classify as decadent, but there is a noticeable tendency in present-day bourgeois art, one which exerts its influence as 'a field of force' in the debate concerning the struggle against this decadence or having recourse to it.

INDEX OF NAMES

Adamov, A., 286
Adorno, Th. W., X, 19, 22, 41, 121, 123, 124, 148, 149, 176, 177, 181, 199
Allport, G. W., 194
Anders, G., 98, 192, 193
Anderson, J., 205
Antonioni, M., 249
Apollinaire, G., 247., 283
Aragon, L., 247, 248, 279
Aristotle, 42, 65, 66, 69, 70—73, 75, 80, 92
Aron, R., 199
Augustine, Saint, 231

Bach, J. S., 114
Bacon, F., IX, 155
Bahr, H., 278
Balzac, H. de, 215
Barthes, R., 262
Bazin, J., 284
Beckett, S., 247, 281
Berkeley, G., 104, 150, 153, 154
Blackstone, W., 129
Bloch, E., XV, 278
Bloch, H., 278
Bourdieu, P., XIII
Bradley, F. H., 34
Brecht, B., 137, 278, 279
Breton, A., 247
Büchner, G., 186

Caillois, R., 244
Calvin, J., 228
Camus, A., 254
Capote, T., 249
Cervantes, S. M., 219
Cézanne, P., 277, 279, 281, 282
Chaplin, C., 192
Childe, G., 40
Churchill, W., 261
Cicourel, A., XII
Cohen, R. S., XVI
Cromwell, O., 212

Danton, G., 186
Daumier, H., 254
Descartes, R., 17, 92, 151, 152, 154
Diderot, D., 246
Dürrenmatt, F., 137, 243, 254, 258, 284, 286

Eckhart, J., 228
Ehrenburg, I. G., 238, 261
Éluard, P., 248
Engels, F., 7, 9, 48, 145, 224, 230
Erasmus, D., 256
Eudoxos, 73

Fetscher, I., 200
Feuerbach, L. A., 201, 225
Fichte, J. G., 17, 111, 151, 156, 159, 160, 176
Fielding, H., 113
Fiore, T., 232
Foucault, M., 266
Fourier, F. M. C., 245, 246, 249, 254
Frisch, M., 259

Gadamer, H.-G., XII, XIII
Gasquet, S. A., 277, 282
Gehlen, A., 53, 125, 126, 170, 176, 202
Geuss, R., XII
Gide, A., 254
Goebbels, J., 261
Goethe, J. W., 1, 4, 31, 46, 65, 76—80
Gogh, T. van, 279
Gogh, V. van, 279
Goldmann, L., 20
Gombrich, E. H., XIII

Goodman, N., XIII
Gorbatov, A. V., 192

Habermas, J., XII, 98, 99, 181, 199
Händel, G. F., 114
Hartmann, N., 48, 50, 59, 81, 92, 133, 140, 177, 179
Hegel, G. W. F., IX, 2—5, 7, 8, 12, 17, 20, 24, 30, 44, 50, 57, 63, 67, 69, 73, 75, 76, 78—80, 83, 86—88, 90, 93, 100, 106, 108, 124, 145—147, 149, 151, 155, 156, 159, 161—164, 179, 180, 182, 200, 256
Heidegger, M., 63, 66
Held, D., XII
Heller, Á., 115, 139, 167, 218
Heller, J., 259
Helmont, J. B. van, 42
Heraclitus, 69, 70
Hess, W., 284
Hochhuth, R., 261
Hoess, R., 261
Horkheimer, M., 19
Huizinga, J., 254
Hume, D., 27, 150, 153—155
Husserl, E., 51, 63, 65—68, 81, 123, 136, 137, 147, 148, 164, 198

Ibsen, H., 263
Illyés, Gy., 271
Ionesco, E., 259, 280, 284

Jaeger, W., 72
James, W., 163, 164
Jauss, H. R., XIII
József, A., 241, 243, 247, 248

Kafka, F., 98, 285
Kant, I., 1, 16, 17, 34, 65, 74—76, 78, 86, 104, 155—157, 170
Károli, G., 268
Kerouac, J., 254
Kierkegaard, S., 123, 254, 257—260
Kirov, S. M., 261
Kołakowski, L., 165, 200, 206
Kopit, A., 259
Kosik, K., 100, 172, 174, 200, 219
Kostov, T., 236

Laertius, D., 73
Landler, J., 120
Lautréamont, 282
Lefèbvre, H., 267
Leibniz, G. W., 16, 92
Lenin, V. I., 7—9, 18, 43
Lessing, G. E., 217
Lévi-Strauss, C., 195, 262
Lipps, Th., 277
Locke, J., 152, 153, 224
Luckmann, T., XII
Lukács, Gy., XVI, 16, 17, 19, 20, 23, 37, 48, 75, 78, 119, 122, 124, 132, 161, 164, 167, 169, 182, 200, 208, 220, 240, 243—245, 267
Luther, M., 211, 232, 268

Malthus, T. R., 205, 211
Mann, T., 244, 254, 258, 277
Mannheim, K., 199, 229
Marcuse, H., 19, 22, 120, 181, 195, 199, 266
Márkus, Gy., 195
Marx, K., 4—7, 13, 16, 18—20, 28, 31, 33, 44, 48, 58, 83, 84, 92, 93, 96, 97, 102, 110—112, 114—118, 122, 124, 126—129, 132, 140, 142, 145, 154, 163, 164, 167, 175, 192, 199, 201, 202, 207—209, 214, 220—222, 225, 249, 250
Mehring, F., 230
Melanchthon, P., 211
Merleau-Ponty, M., 171
Mills, C. W., 131
Mirabeau, H.-G., 212
Modigliani, A., 119, 120
Montaigne, M. E., 254
Mozart, W. A., 245, 246, 254
Münzer, T., 206, 228

Nabokov, V., 254
Nezval, V., 279
Nietzsche, F. W., 67, 68, 170, 277

O'Neill, E. G., 132, 180
Ordjonikidze, G. K., 239, 261

Packard, V., 234
Parsons, T., XIII
Peirce, C. S. S., 163
Pinkus, Th., 169
Pinter, H., 259
Plato, 69—72, 115
Plotinus, 72
Polányi, M., XII
Putna, V. K., 239
Pyrrho, 73

Rauschning, H., 261
Ricardo, D., 161, 205, 221
Robespierre, M.-F.-M.-I., 207, 211, 214, 221, 225
Rolland, R., 113
Rousseau, J.-J., 246

Sartre, J.-P., 20, 48, 54, 63, 67, 68, 124, 215, 254—256, 258, 260, 283
Saussure, F. de, 267
Scheler, M., 63, 147, 148
Schelling, F. W. J., 129, 156—160, 176
Schelsky, H., 212, 213
Schiller, J. C. F., 173, 217, 245, 254, 259, 260, 272
Schlegel, F., 254
Schlick, M., 68, 265
Schütz, A., XI, XII
Sebag, L., 197, 262
Selye, J., 42, 43, 101
Seuse, H., 228
Shakespeare, W., 113
Shaw, G. B., 254
Simmel, J. M., 181
Simonov, K. M., 261
Socrates, 256
Sombart, W., 121
Spinoza, B., 16, 151, 221
Stalin, I. V., 199, 200, 225, 226, 235, 238, 239, 273
Stirner, M., 111, 112

Tati, J., 249
Tauler, J., 228, 232
Thompson, J. B., XII
Tieck, L., 254
Tukhachevsky, M. N., 238, 239
Tzara, T., 282

Updike, J., 249

Vaihinger, H., 69
Vierkant, A., 110
Voltaire, F.-M. A., 177

Waismann, F., 265
Wartofsky, M. W., XIII
Weber, M., 114, 121, 129, 186, 218, 230
Weiss, P., 259
Wilde, O., 277
Wittgenstein, L., 195, 262—266
Wolff, K. H., XVI

Yakir, J. Z., 239

Zhukov, G. K., 261
Zola, É., 196

INDEX OF SUBJECTS

aesthetics, 259, 277
agnosticism, 104, 155
Albigensians, 232
alchemy, 40
alienation, 115, 118, 221, 223, 254, 286; self-, 148
ambiguity, 50, 62, 73, 74, 106, 120, 121, 241, 244, 245, 249, 263—266, 268, 269, 274—275
anima mundi, 2, 17, 151, 200
anthropology, 244
anthropomorphism, 207
anthropomorphization, 51
antinomy, 16, 33, 74, 75, 77; theory of, 156
apparency, 19, 38, 44, 73, 111, 153, 155, 202, 207, 209, 265; theory of, 70
appearance, IX—XIV, 1—15, 20—25, 30, 32—36, 38—42, 44—53, 55, 57, 63—71, 73—81, 84—97, 100—105, 108, 109, 112, 113, 115, 133, 134, 140, 148—150, 154, 156—159, 167, 196, 217, 248, 276, 277; -theory, 21, 22, 66; theory of, XII, 2, 5, 57, 75, 76, 81, 82
art, 4, 124, 215, 275—279, 283, 285, 286; philosophy of, 277
astronomy, 39, 72
atomic theory, 42
avant-gardism, 279, 284, 285

being *(Sein),* 30, 49, 50, 67, 88—90, 95, 100, 104, 106, 107, 112, 125, 130, 140, 146, 157, 163, 173, 258; -as-, 50; chosing, 23; degrees of, 49, 50, 60, 90, 104, 118, 157; for itself, 50, 161; for me, 25, 50; for others, 24—28, 30—33, 45, 46, 61, 64, 84, 94, 96, 97, 99, 146, 149; for us, 24, 35, 45, 49, 50, 57—61, 80, 89, 90, 103, 104, 134—136, 154, 157, 162, 188, 223, 224; form of, 43, 163; human, 5; in itself, 50, 58, 61, 125, 146, 161, 223, 224; mode of, 50, 58, 59; object-, 123; so, 57, 58, 61, 83—85, 97, 106, 119, 129, 135, 171, 195, 218; social, 58; species-, 6, 64, 111, 203, 218, 220, 222, 223, 243; subject-, 21, 38, 43, 44, 46—49, 55, 58—60, 89, 90, 100, 172, 223
biology, 74
Bolshevism, 281

Calvinism, 205
capitalism, 29, 107, 111, 115, 119, 121, 143, 147, 180, 193, 231, 235, 272; history of, 272
(Catholic) Church, 113, 190, 204, 205, 218, 228, 231, 273
causality, 177, 178, 181, 186, 188, 190, 231
chemistry, 32, 74
Christianity, 111, 143
Club of Rome, XIII
cognition, 2, 6—10, 14, 21, 28, 33, 35—39, 41—45, 48, 49, 54, 56, 57, 59—62, 66, 69, 71, 72, 77, 79, 83, 89, 99, 100, 101, 104—107, 151, 152, 158, 162, 163, 167, 194
colour-theory, 76
communication-theory, XI
communicative action: theory of, XII
communism, 284
conscience: guilty, 144, 145, 226
consciousness, X, XV, 8, 9, 14, 17, 19—25, 31, 33, 35, 38, 40, 43, 46, 51, 54, 55, 58, 60, 61, 83—85, 88—90, 92, 93, 95, 97, 99—103, 105, 106, 109, 111, 116, 118, 120, 123, 125, 126, 129, 131, 133, 136, 137, 141, 145, 146, 149, 150, 161, 162, 172, 173, 175, 176, 179, 184—193, 197—200, 203, 214, 218, 223, 224, 226, 227, 229, 230, 234, 236—240, 244, 258, 265, 267, 268, 272, 273; background, 238—240; class-, 94, 185; everyday, 87, 93, 127, 133, 168, 171, 176, 191, 192, 197, 200; false, XII, 9, 30, 45, 47, 89,

(consciousness continued)
94, 102, 113, 114, 149, 165, 170, 185, 188, 191, 202, 210, 222, 224, 226, 227, 229—231, 233, 234, 236—240, 265; theory of false, XII; form of, 17, 87, 147, 153, 163, 167, 178, 181, 188, 189, 191, 219, 223, 226, 227, 231, 233, 234, 239, 268; formations of, 162; practical, 37, 126, 132, 134, 140, 152, 153, 167, 168, 170, 172, 174—176, 179, 192, 197, 198, 216; pure, XII; social, XIII, 10, 47, 53, 64, 85, 173, 174, 177, 181, 240, 263, 268; teleological, 52, 126, 169, 172, 177
contradictions: theory of, 159
Cubism, 281, 282
cybernetics, 194

Dadaism, 247
determinateness, 12, 13, 106, 132, 140; contentual, 66, 67; reflectional, 7, 31; reflexive, 44, 93, 94; relational, 4; social, 132; teleological, 95, 132
dialectic, 1, 7, 10, 16, 18—20, 27, 33, 35, 48, 57—59, 68—70, 75, 96, 97, 116, 141, 145, 146, 157, 159, 212, 223, 255, 258, 280, 281; of appearance and essence, 6, 8, 45, 47, 66, 75, 78, 81
dialectics, 19, 65, 69, 75
disanthropomorphization, 177
disideologization: theory of, 240
dogmatism, 20, 27, 170, 181, 199, 201, 228, 229, 235, 237, 276

economics: history of, 231; political, XIII; vulgar, 221
economy, 203, 212, 234, 241; political, 154
Eleatics, 69, 70, 72
empathy: theory of, 277
empiricism, 151, 152, 156
Enlightenment, XIV, 121, 127, 199, 201, 207, 217, 224, 225, 245, 253, 254; British, 34, 150, 152, 155
epiphenomenon, 68
epistemology, 1, 3, 8, 30, 36, 43, 45, 46, 57, 60, 62, 86, 123, 140, 163
essence, 1—11, 13, 30—32, 36, 44—49, 63, 66—69, 71, 74—81, 86, 88, 89, 99, 102, 116, 146, 147, 159, 161, 172, 192, 197, 218, 252, 267, 276, 282, 283, 285
ethic, 143, 144, 188, 205, 210, 228, 230
ethics, 265, 272
ethnology, 195
ethnophilosophy, 262
ethos, 144, 212
existence: mode of, 13, 22, 32, 39, 54, 78, 80, 83, 86, 88, 90, 94, 106, 108, 109, 111, 121, 141, 148, 153, 157, 159, 181, 190, 202, 205, 224, 225, 232
existentialism, 35, 254
externalization, 46, 47

factuality, 48, 62, 119, 120—123; theory of, 123
φαινομενα, 69, 70
fascism, 261, 280, 281, 284
fetishism, 96
fetishization, 122, 123, 134, 137, 238, 275; theory of fetishization and objectification, 22
fetish-theory, 18, 83
feudalism, 113, 121, 144, 174
forms: doctrine of, 64; theory of, 92
Frankfurt School, XII, 12, 19, 22
freedom: realm of, 249, 250; theory of, 259
free fall: theory of, 42

games: theory of, 244, 253, 259
generality, 96, 141, 148, 161, 202, 203, 208, 209, 217, 221
generalization, 68, 87, 193, 211, 256, 261
geometry, 15
gnostics, 232

Hegelianism, 200
hermeneutics, XIII
Hinduism, 210
historicism, 190, 225
historicity, 2, 6, 33, 120, 174, 190
historiography, 190
history, XIII, XIV, XVI, 2, 6, 17, 18, 64, 81, 106, 112, 119—121, 143, 162, 200, 210, 219, 222, 229, 231—234; cultural, XIV; cunning of, 237; philosophy of, 159, 253;
homogeneous medium, 16, 25, 47

idealism, 64, 79, 104, 151, 155, 159, 165; German, 155; objective, 157, 158; subjective, 104, 150, 153, 155, 159

Ideas: doctrine of, 69—72
ideology, 48, 74, 102, 188—191, 199—230, 234, 236, 238, 240, 248, 262, 267—270, 275; theory of, 188, 199, 201, 225; theory of freedom from, 199
illusion, IX, 5, 6, 20, 21, 25, 28, 29, 34, 38, 39, 41, 45, 47—49, 53, 54, 57, 59, 62—68, 71, 74, 78, 80, 81, 83—92, 95—99, 101—105, 107—110, 114—116, 118, 119, 121, 122, 124, 127, 129, 130, 132, 139, 140, 143, 146—150, 152, 154, 156, 159, 160, 163, 164, 167, 169, 171, 175, 176, 179, 180, 187, 189, 190—192, 194, 195, 198, 201, 206, 207, 211, 213—219, 222, 226, 232, 234, 235, 237—239, 241—245, 248—258, 260, 262, 268, 271, 272, 274—280, 282—285; doctrine of false, 80; history of, 243; theory of, 111; theory of social, 62
illusoriness, 44, 69, 111, 151, 153—155, 220, 223, 241—244, 258, 259, 276—279, 284
immediacy, 16, 20, 28, 34, 46, 47, 49, 54, 55, 86—88, 92, 100, 103—117, 119, 124, 129, 133, 147, 165, 172, 175, 211, 222, 243, 246, 257, 269—271, 278—280, 283, 284; theory of, 111
imperialism, 215
impressionism, 282
irony: theory of, 259
irrationalism, 75, 111, 156—158
irrationality, 265

Jacobins, 209
jurisprudence, 129

knowing subject, 3, 10, 17, 21, 24, 34, 49, 51
knowledge: sociology of, XIV

land rent: theory of, 210
language: philosophy of, XI, 194
law, XIII—XVI, 13, 14, 20, 22, 23, 30, 33, 34, 74—76, 79, 80, 89, 97, 102, 108, 140, 146, 160, 191, 203, 210, 212, 213, 223, 252; biological, 212; bourgeois, 122; Common, 129; economic, 234; of economics, XIV, 193, 234; feudal, 122; natural, 52, 109; theory of natural, 221; of nature, 192, 223; precedential, 130; social, 212
linguistics, 69, 195, 265—267
literature, 196, 245, 258, 281
logic, XI, 35, 94, 113, 142, 148, 160, 167—170, 176—178, 181, 187, 198, 199, 209, 210, 225, 231, 239, 262, 273, 276; mathematical, 194, 195, 197
Luddism, 240

manipulation, 83, 88, 96, 100, 118, 120, 121, 176, 215, 219, 224, 234—238, 240, 241, 243—245, 265, 266, 269, 275; gross, 200, 235, 240, 273; subtle, XVI
Marxism, 5, 7, 19, 33, 68, 75, 82, 164, 225—227, 238, 240, 270
Marxist theory, 230
materialism, 8, 58, 165, 200, 211; dialectical, 27; vulgar, 27
matter: and mind, 43
Middle Ages, 64, 143, 144, 190, 204, 205, 208
monetarist theory, XIII
mysticism, 208, 228, 156

naturalism, 196
nature: philosophy of, 158
necessity: realm of, 249, 250
neo-Hegelians, 111
neopositivism, 68, 69, 120, 122, 123, 150, 165
noumenality, 50, 100
noumenon, 17, 49, 68, 74, 78

objectification, 3, 16, 22, 29, 59, 60, 75, 85, 93, 117—119, 127, 128, 138, 150, 163, 167, 175, 176, 181, 187—189, 191, 195, 242, 250—252; -theory, 22
objectivity, 2, 3, 5, 8, 16—20, 22—24, 28, 31, 38, 43, 44, 49, 59—61, 64, 66, 67, 69, 72, 73, 75, 80, 83, 85, 86, 88, 96—98, 103, 105, 118, 121—124, 126—132, 134—137, 139—143, 145, 146, 148, 149, 168, 173, 174, 179, 186—190, 223, 242, 250—252; indefinite, 267; objectless, 173

object-making, 59, 124—142, 148, 149, 163—165, 252
ontological theory, 19
ontology, 3, 5, 9, 17—20, 28, 30, 36, 38, 39, 46, 47, 50, 57, 62, 69, 78, 92, 177; action, 176; formal, 92; social, 31, 34, 213; social and historical, 19; of society, 18; socio-, 6; theory of, 9
operationalism, 195

particularity, XIII, 27, 68, 78, 80, 106—108, 112, 116, 150, 157, 158, 202, 203, 226, 234, 237
phenomenality, 57, 277
phenomenology, 5, 24, 35, 47, 48, 51, 56, 57, 59, 65, 67—69, 73, 81, 82, 121, 122, 137, 148—150, 171
phenomenon, IX, XIII, 1, 3, 6—16, 20—32, 35—39, 41—44, 46, 47, 49—58, 60—62, 64, 66, 67, 69, 70, 72, 74, 77, 81, 82, 84, 89, 90, 92, 96, 99, 101, 111, 136, 143, 148, 150, 151, 153, 154, 163, 177, 178, 180, 196, 198—200, 204, 216—218, 223, 225, 227, 236, 238, 239, 241, 244, 246, 256, 258, 261, 263, 264, 266, 278—286; natural, 2—5, 7, 17, 24, 31, 45, 91; social, 3, 20, 23, 24, 27, 31, 45, 46, 48, 51, 84, 91, 96
'phenomenon', 20, 21, 34—38, 40—58, 60—67, 69—74, 76, 78—82, 86, 88, 92, 100, 101, 127, 130, 135, 148, 155, 280; theory of, 74, 75, 80
philosophy, 2, 4, 9, 18, 19, 22, 31, 35, 52, 53, 57, 62, 64, 66—69, 79, 80, 150, 151, 154—156, 164, 170, 197, 221, 245, 270; bourgeois, 34, 48, 68, 197; British, 159; classical, 37; European, IX; (classical) German, XII, 1, 4, 22, 65; Greek, 65, 70, 72, 73, 114; Hegelian, 245; historical, 245; history of, 16, 20, 35, 37, 43, 44, 50, 64, 66, 75, 80, 92, 111, 150, 155; Kantian, 56; linguistic, 137, 197; Marxist, 181; modern, 34, 37, 48, 62, 92, 120, 150, 165, 197, 201; moral, 211; natural, 1, 2, 3, 6, 7, 9, 18, 31, 34, 35, 46, 47, 49, 55, 62, 65, 72, 73, 76, 156
physiocrats, 211
play: theory of, 250, 253
politics, 224, 241, 285
positivism, 154, 165
practice, X, XI, 1, 3, 6, 9, 10, 17, 21, 36, 38—40, 43, 44, 47, 49, 51, 53, 54, 56, 59, 60, 61, 83, 89—92, 94, 97, 100, 101, 105, 108, 110, 111, 117, 118, 120, 122, 124—131, 132, 134, 135, 140, 141, 149, 155, 163, 164, 167, 169, 171—177, 180, 187, 189, 193, 194, 201, 202, 204—214, 216, 223, 225, 227, 230—233, 236, 250—252, 254, 260, 267, 272, 276; theory of, 44
pragmatism, 69, 163, 165
praxis, 1—4, 6, 7, 9—11, 16—20, 22, 24, 25, 27, 31, 36—39, 42, 44—46, 49, 53, 55—61, 64, 81, 82, 89, 91, 94, 97, 99, 100, 102—104, 106, 109, 110, 112, 117, 118, 124, 129, 134, 135, 140, 147, 149, 153, 159, 163—165, 167—170, 174, 188—190, 193, 201, 204, 205, 207, 210—213, 221, 223, 226, 227, 233, 237, 267, 272; theory of, 164
predestination: theory of, 231
presentation, 1—3, 6, 9, 10, 14—16, 21—23, 31—34, 48, 50, 52, 58, 61—64, 71, 80, 81, 83—85, 91, 99, 106, 149, 157, 217; form of, 2, 6, 7,14, 16, 23, 32, 46, 49, 55, 58, 60, 61, 65, 84, 92, 93, 96, 100, 101, 116, 119, 131, 134, 143, 146—149, 152, 167, 175, 189, 205, 206, 212, 216, 218, 219, 222, 224, 236, 241, 250, 262; theory of, 1, 48
presentational form, 32, 62, 63, 81, 83, 86, 89—92, 96, 100, 106, 159, 167, 171, 175
primary and secondary qualities, 152; theory of, 34, 153
prime mover: theory of, 72
psychoanalysis, 244
psychology, 132, 226, 244
Puritanism, 212, 230
Puritans, 204, 230
Pythagoreans, 70

rationalism, 150—152, 156, 207, 233
rationality, 160, 179, 240
realism, 64, 254; socialist, 279
reality, X, XII, XIV, XVI, 2, 3, 7, 10, 11, 16, 17, 21, 24, 25, 28, 29, 36, 38, 39,

(reality continued)
41—45, 48—51, 53—56, 58, 59, 62, 64, 66, 69—71, 74, 76, 79, 80, 83, 85, 86, 88, 89, 92, 98, 104, 107, 109, 112, 118, 119, 122—125, 129, 130, 135—137, 144, 145, 148—158, 160, 163—165, 167—169, 171, 172, 174, 175, 178, 180, 189, 190, 193—195, 198, 199, 201—210, 213—220, 223, 225—227, 229, 231, 232, 233, 236—239, 241—245, 250—254, 257, 260, 269, 271, 274—285; pseudo-, XI, XV, XVI; quasi-, 215, 217
reason: cunning of, 105, 162, 169; practical, 156; religion of, 208; theoretical, 156
reflection, 7, 14, 16, 30, 31, 43—45, 61, 65—67, 70, 75, 79, 84—86, 90, 106, 107, 118, 119, 121, 122, 126, 127, 129, 145, 158, 163, 167, 188, 189, 207, 229, 230, 276, 278
Reformation, 208, 233; German, 227
reification, 119, 127, 148, 149
religion: critique of, 199; sociology of, 173
Renaissance: Spanish, 217
revolution: beorgeois, 159, 208; English, 212, 224; French, 208, 210, 214, 225, 245; glorious, 224; industrial, 155; scientific, 17, 74; technological, 25, 128, 199, 245
roles: theory of, 254
Romanticism, 256

salvation: theory of, 204
scepticism, X, XI, 7, 153—155, 219, 254, 262
scholasticism, 177
science, 4, 6, 17, 18, 32, 36, 42, 47, 62, 63, 65, 68, 72, 74, 81, 84, 100—102, 148, 164, 169, 196, 197, 234, 249, 262; development of, 41, 48, 100, 264; exact, 5, 18, 57, 74, 177; experimental, 249; factual, 123; history of, 88, medical, 40, 42; natural, 4, 47, 51, 74—78; philosophy of, XI, XV; positive, 81; progress of, 7, 38, 61, 74, 101; social, XI
Second International, 226, 230
semantics, 69
semiotics, 262, 266, 269
singularity, 68
social criticism, 176, 199, 245, 249, 278
Social Democracy, 225
sociology, XI, XII, XV, 47, 98, 181, 199, 229, 261, 262
Stalinism, 273
structuralism, 69, 194, 197, 262, 266, 267
subject and object, 17, 19, 37, 58, 59, 60, 179; identity of, 18; unity of, 19—22, 37, 39, 43—47, 77, 110
subjectification, 18, 67, 167
subjectivism, 207, 257
subjectivity, 9, 18, 22, 23, 31, 36, 51, 52, 73, 75, 81, 85, 86, 102, 103, 150, 160, 176, 179, 196, 258
substance, 70—72, 91, 133, 151, 154, 156, 164, 192—194, 206, 237, 248
Surrealism, 247, 284

teleology, 22, 56, 170, 179, 182, 183, 186, 187, 190
thermodynamics, 33, 74
thing: -in-itself, 1, 7, 140, 173; itself, 143, 145, 147, 149
transmigration of souls: doctrine of, 210
truth, X, XII, XIV, XV, 51, 55, 69—71, 82, 86, 88, 91, 102, 103, 147, 151, 152, 155, 161, 162—164, 165, 172, 174, 175, 195, 198, 201, 206, 207, 209, 212, 215, 216, 220, 224, 232, 233, 241, 261, 262, 271; -content, 35, 258; half-, 270; pseudo-, XII; semi-, XIV

universality, 24, 68, 77, 78, 131, 180, 183, 218
universe: geocentric theory of, 211
unreality, 150, 151, 225, 244, 247, 251, 252, 257, 259, 269, 274
untruth, X, XII, 2, 85, 153, 161, 262, 263, 275, 276
Urphänomen, 77, 79, 80; theory of, 76, 78, 80
utopia, XV, 99, 229, 245, 249, 263

value: labour theory of, 161; theory of, XIII, 63; theory of surplus, 107
verisimilitude, 38, 153, 225
Vienna Circle, 63, 68

world picture: anthropocentric, 177; mechanical, 43; modern, 197; scientific, 206, 283